**IEE MANUFACTURING SERIES 1**

Series Editor: A. Davies

# Developing a

# MAKE or BUY

## strategy for manufacturing business

Developing a

# MAKE or BUY

## strategy for manufacturing business

## DAVID PROBERT

The Institution of Electrical Engineers

Published by: The Institution of Electrical Engineers, London,
United Kingdom

© 1997: The Institution of Electrical Engineers

The Institution of Electrical Engineers,
Michael Faraday House,
Six Hills Way, Stevenage,
Herts. SG1 2AY, United Kingdom

**British Library Cataloguing in Publication Data**

A CIP catalogue record for this book
is available from the British Library

**ISBN 0 85296 863 9**

# Contents

# Acknowledgments

The development of the make or buy methodology has been in collaboration with, and supported by, many individuals. Encouragement, guidance and support came from Charles McCaskie of the Royal Academy of Engineering, and Mike Gregory, Professor of Manufacturing Engineering at the University of Cambridge.

Several members of the Manufacturing Business Unit of CSC Computer Sciences Limited contributed to the development and subsequent assessment of the methodology, through pilot studies within Lucas Industries plc and at a number of external companies. Much of this work was jointly coordinated through John Davies, a former member of Lucas Engineering & Systems, in his role as principal consultant with CSC Computer Sciences Ltd. The author and publishers wish to acknowledge his and the other individuals' contributions to the development of the methodology, and the permission to use and adapt various materials.

# Foreword

The issues surrounding the make or buy question in manufacturing industry
· are as old as the manufacturing activity itself, and interest in finding a sound
way of dealing with them has been a constant feature of industrial life.

This book represents the results of a research and application project
carried out by the author while an industrial research fellow, based in the
Manufacturing Engineering Group at the University of Cambridge, and
supported by Lucas Industries and the Royal Academy of Engineering.

The area of application was of immediate interest to Lucas, which faced
choices of where to concentrate its manufacturing capability. Development
and application of the ideas inside Lucas (and other) businesses was carried
out in collaboration with Lucas Engineering Systems, at that time the Lucas
consulting business.

As an area for research, make or buy issues are a central part of the
manufacturing strategy of a business, which is a core research theme at the
Cambridge Manufacturing Group. Other Group research interests include
the management of technology, new product introduction, performance
measurement and international manufacturing. All programmes are carried
out with the active involvement of manufacturing industry, with the aim of
transferring practical new approaches to dealing with these issues to
participating companies and beyond.

The author would like to thank all those in the Manufacturing
Engineering Group who have supported the work and contributed to the
development of the ideas, LE & S friends and colleagues who enthusiastically
applied and developed the approach, Lucas businesses for providing access
and support to test and apply the approach, in some cases before it was
completed, and the Royal Academy of Engineering for continued support
and encouragement.

I hope that the ideas and methods discussed here will be of help to many
other manufacturing businesses as they make choices about their
manufacturing capability. The team based approach to formulating a new
make or buy strategy described in this book has been applied in many
different businesses and provides a clear and practical route to improving
manufacturing performance.

*Chapter 1*

# Introduction

## 1.1 Origin and purpose of this book

The contents of this book are based on the results of research, application and development carried out in collaboration between university and industry. The idea was to provide a soundly based practical approach that would enable people in manufacturing businesses to develop a make or buy strategy tailored to the requirements of their own company.

During the course of the project a number of publications were produced to give an account of the development of the ideas and to enable wider discussion and criticism of the approach. This has been through the medium of academic journal and conference papers, and a short management guide published by the Department of Trade and Industry under the Managing in the 90s information and awareness progamme. Entitled 'Make or buy: your route to improved manufacturing performance?'[1], this booklet gives an overview of the issues involved in make or buy considerations. The bibliography gives a list of these and other relevant publications.

The intention from the start of the project was to produce a book giving a fuller account of the issues involved in make or buy considerations, including other contributions to the subject from the perspectives of manufacturing engineering, economics, accounting and other relevant disciplines. However, most importantly, the book should contain a practical guide to devising your own make or buy strategy, which could be applied by managers facing these issues for real in their own company. This book is the result of that intention and I hope that it enables teams of people in manufacturing industry to work together in understanding and exploiting the possibilities that face them in a competitive world.

## 1.2 What is make or buy all about?

Fundamentally make or buy is about the choice of whether to carry out a particular process or activity within your own business or to buy it in from a supplier. In reality this can take many forms: choice about making a particular small part of a complex larger product (for example, if we make turbine generators, should we make the ball bearings that go into them?), choices about system and subsystem manufacture (if we make cars, should we make the engine?), choices about which particular manufacturing processses to have in the company (if our manufacturing process requires a specialised heat treatment, should we buy the plant or use a specialised subcontractor?). Although the examples given are from manufacturing businesses, the same issues apply to other in or outsourcing decisions. For example, should we, as a business, provide our own catering, security and data processing facilities, or should we contract these out to specialised service providers? These issues are of equal relevance in manufacturing and service environments, including (we have found) the health and prison services! The distinctive feature of manufacturing industry is the variety of processes involved and hence the frequency with which the issue is encountered, with the resulting greater scope for good or bad decisions to affect the business result.

The importance of the subject arises not out of each individual decision, but from the fact that, over a period of time the consequences of all the decisions actually determine the size and nature of the whole enterprise. In a sense, the boundary of the business is the fundamental consideration at the heart of make or buy, and the level of vertical integration of a company, or the range of technologically different activities conducted within its boundaries, will be the result of many past make or buy decisions.

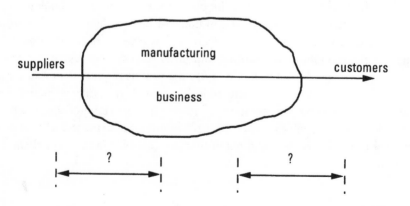

*Fig.1.1 Make or buy is concerned with the strategic issue of where to position the boundaries of the business*

## 1.3 What drives a company to tackle make or buy?

Very often companies find themselves in the area of make or buy decisions without that being the original intention. Efforts to be more selective about the number and quality of suppliers, debate about where the company should really concentrate its resources, or an attempt to define the core competences of the organisation, can all lead to the need to evaluate which activities should be carried out inside the company and which outside.

At the same time there may be an awareness that successful and profitable companies may have very different approaches to the range of activities which they carry out in house or subcontract out, depending on the industry sector they are in and what is emerging as the current successful form for the enterprise. Historically, this form has evolved and changed over the decades. It is dependent on the available range of capability that is present among potential suppliers, and the overall state of the economy.

Experience during the course of this work has covered both downturns and upturns in the national economy, with companies seeking either to expand or contract. In both environments issues central to make or buy considerations were found to be at the forefront of management thinking. During recessionary times there may be a requirement to reduce the fixed cost base of the business, or simply costs in general. This needs to be done without prejudicing the long term viability or essential manufacturing capability of the company. On the other hand, when the economy is expanding, the company may be looking for ways to increase capacity rapidly and either not have the capital to finance this to the full internal extent, or be unwilling to take the risk of long term commitment to particular capabilities. In all cases, the need for a rigorous assessment of all the variables affecting the decision is apparent.

Managers are much influenced by the structure of the obviously successful companies in their field. The car industry is a good example of this, where over the years the model of successful enterprise has evolved and changed. In the early days the Ford company was involved in nearly every stage of the manufacturing process, even as far back as sheep farming to produce wool for seat covers. A very successful and profitable business was built on this high level of vertical integration, until General Motors (in its then guise) found a more flexible and competitive business structure, which gave more freedom to suppliers to innovate. In recent years car companies have become steadily less vertically integrated, while retaining control of design and systems integration aspects of the finished product, together with assembly and testing. The dilemma is that while most companies follow the trend set by the organisational form of the emerging successful competitor, in order not to miss some advantage that this appears to give, it is the company which establishes the innovative new form of organisation or new way of working that will become the market leader. In consequence, the enlightened and ambitious company will be constantly examining its options for the most advantageous level of vertical integration, i.e. the range of activities which it should carry out within its own business boundaries. The implication for the need for a systematic way to develop and assess these options is obvious, given the benefits of a structured approach that can be recorded and revisited as conditions change.

## 1.4 How has industry tackled the subject up to now?

Given the perennial relevance of the subject, there is surprisingly little record of a systematic approach to integrating all the issues that need to be considered simultaneously in coming to conclusions about make or buy. Engineering and management literature over past decades shows that interest in the subject has remained consistently high, but that it is usually tackled from a fairly narrow cost-based perspective. There have been many contributions from authors with a finance or accounting background, exploring aspects such as the impact of different costing systems, payback calculation and so on.

An early study of the issues involved was published in 1942 by Professor Culliton of Harvard University[2]. He carried out a case based review of companies in the eastern part of the United States operating in the first half of this century, and grouped the factors into three categories: cost, quality and quantity. He felt compelled to come to a conclusion ( which was that on balance buying is to be preferred to making !), but more significantly makes the very important observation that the result of a make or buy review is dependent on the current economic, social and political context. In consequence, the conclusions of such a review should be periodically revisited. More recent treatments of the subject have emphasised additional factors, as will be seen in the following chapter. The most significant among these are aspects of inter firm relationships, the ideas of competitive strategy and strategic management and the increasing importance of technological capability and resource.

Another useful study from the accounting perspective was produced by Anthony Gambino in 1980 on behalf of the National Association of Accountants in the USA and the Society of Management Accountants of Canada[3]. As you might expect, this work concentrates very much on the financial aspects of make or buy. This is an area (like most others in the management arena) where the focus of interest shifts over the years, and to some extent fashions come and go. Gambino's book gives a full account of investment justification and the application of payback calculation. In addition it also includes the results of a survey of practice among participating firms, and highlights a number of useful and relevant points. The multidisciplinary nature of the issue is discussed, with the resulting need for participation from many different functions within the company in coming to conclusions. The role of committees or teams set up to carry out make or buy decisions in a consistent way is described, and the value of having some kind of decision path or chart to follow is demonstrated.

Some particular additional aspects are developed, such as the effect of activity level on costs, and the dependability of suppliers, but there is no recommendation as to how all these factors should be brought together in a comprehensive way. An emerging difficulty is the integration of the comparatively easily calculated factors (cost based) with the rather more subjective considerations (future activity levels, actions of customers, suppliers and competitors, etc.) into one consistent review process. This is the area where I hope this book makes a contribution, in particular from the perspective of the manufacturing manager.

More recently there have been other surveys of industrial practice and interest (including my own, of which more later). In particular the aerospace sector in the UK, under the auspices of the SBAC (Society of British Aerospace Companies), has published a report of good practice based on survey work among its member companies[4]. The importance of a clear policy statement (in effect, a strategy) is emphasised, together with the criteria that need to be weighed up in coming to individual decisions. These criteria are grouped under seven headings: value (costs), technical, resources, programme (volume, timing), quality, suppliers and strategic factors. The report provides an assessment chart for weighing up these factors in coming to a decision and stresses the importance of multifunction input to the decision process. Finally, two example decision processes, in the form of route maps in use at SBAC member companies, are shown. The formulation of the strategy itself is not an area that is developed in the report, and that is the main purpose of approach described in this book.

## 1.5 What do manufacturing managers want to know about make or buy?

During the course of the project a short survey of large UK manufacturing companies was carried out to determine their current attitude and approach to the issues of make or buy. The purpose of this was that, in the course of the research, project work and preparation of this book, their interests and concerns could be taken into account. If this was not done there was a danger that the particular approach that we were developing would only be of interest to ourselves and the businesses within which we developed and applied the ideas.

The survey contacted about 60 large (over 1000 employees and £50 million turnover) companies across manufacturing industry, covering the automotive, aerospace, electrical, electronic and mechanical engineering sectors. Originally, this was intended to be a pilot survey which would be followed by an amended (if necessary) survey to about 400 companies (approximately the full population of companies of this size in the UK). However, the response to the pilot (about 60%) was so good that it provided sufficient information for our purposes in determining mutual areas of interest and current attitudes. The main points emerging were:

- all companies recognise the need to treat make or buy as part of the business strategy;
- only 50% of companies have defined policies and procedures to support managers in their make or buy decision making;
- only about an third of companies have committed resources to making sure that expertise is available to carry out make or buy decisions;
- an overwhelming majority (85%) would like to have a documented make or buy support framework.

Although the long term impact of make or buy decisions is recognised, many companies still take a much too short term view of the consequences. About half of the responding companies thought in terms of a one year horizon, when three to five years would be more realistic.

The most significant questions that managers wish to be able to answer in coming to make or buy choices is 'What are our key company strengths and how can we exploit and protect them?'.

Although a wide range of factors is recognised as having an effect on the make or buy decision, the issue of matching company capability to what is really required to be successful in a particular business is at the core of a good make or buy strategy. The central plank of a good practice approach to make or buy is the means to make this strategic assessment, and is the main content of this book.

The survey also investigated the different forms of relationship which existed between the responding companies and their suppliers. This will be discussed in more detail in Chapter 3.

## 1.6 Structure of the book

Following this introduction, the next chapter deals with the way make or buy issues fit into the overall context of a manufacturing business. The different levels at which make or buy can be operationalised are reviewed, and the various factors which need to the integrated into the review process are considered.

The contribution of relevant ideas and perspectives from subject areas other than manufacturing engineering are discussed, with a view to seeing how they may contribute to the process. The need for the process itself to be structured is examined, together with the implications for multidisciplinary team working and project management.

Chapter 3 outlines the principle factors and steps to be worked through in carrying out a strategic review of manufacturing operations leading to a new make or buy policy or strategy. The nature of the review process and the resulting policies and procedures which can be implemented to execute the new strategy are described, together with the requirement for maintaining the strategy over time.

Chapter 4 is almost a book within the book. It contains the step by step guide to how a project team can work through the process of devising the new make or buy strategy. Divided into sections, each describing a step in the process, it gives practical guidance, together with tools and techniques, hints and tips, based on real project experience.

Chapter 5 describes the training and project management that is necessary to ensure the successful development of the new strategy and its implementation. In addition, it discusses the need to keep the strategy up to date and outlines the means by which this can be done.

Chapter 6 illustrates the process with examples taken from actual case experience. The different aspects which emerged as important in the various business environments are discussed.

Chapter 7 provides more detail on some of the analytical tools and techniques which are useful in the strategic review. The choice of which to use depends on the circumstances of the particular business under review, and so a variety of possible techniques is presented.

Chapter 8 indicates how a company can go about starting up the process, and describes some of conditions which need to be in place before beginning.

## Chapter 2

# Make or buy in the manufacturing business context

## 2.1 The position of make or buy in business strategy

Make or buy issues sit firmly at the centre of the manufacturing strategy of a company. Although the ideas of business strategy have been developed, discussed and applied by academics and practitioners over many years, the concepts of manufacturing strategy are comparatively recent. The history and content of manufacturing strategy will be examined in more detail, but first we need to see how it fits into overall business strategy.

## 2.2 Manufacturing and business strategy

For most companies business strategy starts by having a top level definition of what the business is aiming for. This usually takes the form of a company mission statement, possibly followed by some specific goals. The mission statement will contain a definition of the service which the company provides for customers, the market position it occupies or wishes to attain and something about the values or culture of the company. Although such statements are very often greeted with scepticism when first introduced, even by the firm's own employees, they are a very important part of direction setting for the business. Indeed, if the top management of the company is unable to articulate this sort of statement or vision for the current and future role of the company, then it will be very difficult to align subsequent actions and engage employee enthusiasm in any coordinated way, particularly in large concerns.

In recent years many companies, large and small, have introduced mission statements and accompanying goal definitions. Clearly this is not sufficient to ensure that the business is thereafter successful, but it is a very important building block for subsequent strategy development. Some consistently successful companies have had such statements, particularly of company values, for many years. If consistently held over time, this has the effect of helping to build a company culture in which individuals know almost instinctively what the company is setting out to achieve and how it wants to operate. Marks and Spencer and Mars could be considered examples of this.

In complex organisations the strategy will now break down into the different business units or operating divisions which make up the whole company. Fundamentally, strategy is about how the organisation makes choices and decisions in order to follow the direction set out in the mission statement. In most companies this will require decision guidelines not only for each business unit, but for the functional areas of expertise within each business that have some autonomy but need to be managed coherently with the others. These strategy elements could cover marketing, research and development, finance, sales and support and, most importantly from the make or buy perspective, manufacturing. The hierarchical relationship of the corporate mission and goals, business unit strategies and strategy elements is shown in Fig. 2.1.

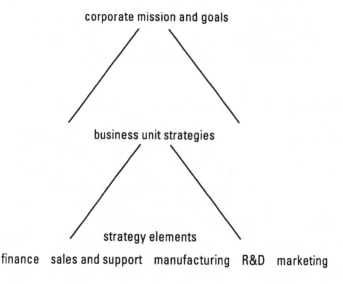

*Fig. 2.1  The link between manufacturing strategy and corporate mission and goals*

## 2.3 Manufacturing strategy

The explicit definition of the decision areas within manufacturing strategy has only developed since the mid 1970s. Until then companies appeared to have thought that strategy was needed for finance and marketing, for

*Table 2.1 The evolution of the decision areas in manufacturing strategy*

| Skinner (1974) | Hayes and Wheelwright (1984) [16] | Hayes, Wheelwright and Clark (1988) |
|---|---|---|
| plant and equipment (span of process) | capacity | capacity |
| | facilities | facilities |
| | technology | production equipment and systems |
| | vertical integration | internal/external sourcing (make or buy) |
| labour and staffing | workforce | human resource policies |
| | quality | quality systems |
| production planning and control | production planning/ material control | production planning |
| organisation and management | organisation | organisation |
| product design/ engineering | | new product development |
| | | performance measurement systems (and capital allocation) |

example, but that manufacturing required little strategic thought and would be able to deliver whatever was necessary. In the late 1960s and early 1970s, as the manufacturing competitiveness of western companies began to slip, a new realisation dawned. A pioneer in this field was Wickham Skinner of Harvard University, whose 1974 article 'Manufacturing: missing link in corporate strategy'[5] established a new way of integrating manufacturing decisions into the strategic management of the business. In principle this amounts to defining the decision areas within manufacturing that need to be considered coherently if manufacturing is going to be able to provide a competitive advantage.

In the years since 1974 these decision areas have been developed and modified, although they are traceable to Skinner's original five areas. Currently we work with the development of these areas proposed by Hayes, Wheelwright and Clark in 1988[6] which provides a little more detail about individual decision areas and so is more practical to apply.

In order for a company to claim to have a manufacturing strategy, it needs to have articulated how decisions are to made in all these areas in a coherent way. Coherent in this context means that the decisions in the different areas are mutually supportive, and that they are in line with overall business and corporate strategy. Further, it is necessary for the manufacturing organisation to have some specific objectives, derived from the business objectives. Manufacturing objectives are usually variations on the theme of quality, cost, delivery and possibly flexibility. Flexibility is considered a possibility because it may not be quoted as a separate objective, but rather as the variations that are required in the other three.

The importance of these objectives is that they are the tangible embodiment of the goal of the strategy and so may be used to assess the alignment of individual decisions with the strategy. However, even more importantly, they are directly linkable to the reasons why customers buy the product, and are variously referred to as key success factors, order winning criteria, critical success factors, etc. in the business literature. We make use of them as a central element in the development of a make or buy strategy.

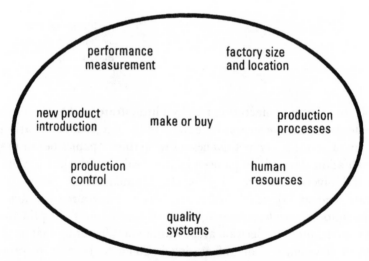

*Fig. 2.2  Make or buy is at the centre of a company's manufacturing strategy*

The key positioning of make or buy in manufacturing strategy is shown in Fig. 2.2. It is one of the structural decision areas, i.e. it has a determining effect on the physical assets, the size of plant and equipment, which are present in the business. In addition, it will have a significant impact on our ability to develop and introduce new products. In many cases, the design knowledge for a new product is dependent on the experience of having made the previous product. For this reason it is the decision area that needs to be addressed very early in the process of formulating a manufacturing strategy.

## 2.4 Levels at which make or buy needs to be considered

There are three ways in which people in manufacturing industry encounter the need to make these decisions.

### *Strategic make or buy*

The strategic approach provides the rationale for investment in manufacturing capability in the long term. Central to the manufacturing strategy of a business, it aligns the choice of which parts of the product to make and which manufacturing processes to have in house with the goals of the business. It provides the framework for the shorter term tactical and component decisions.

### *Tactical make or buy*

This deals with the issue of temporary capacity imbalance. When unforeseen changes in demand happen it may not be possible to make everything in house even though this would be the preferred option. Conversely, if load falls the company may wish to bring in house some work which had previously been outsourced, without damaging important supplier relationships. In this situation managers need a way to choose between the options open to them, within the guidelines of the strategy, and usually on the basis of optimal financial contribution.

### *Component make or buy*

Usually at the design stage, component make or buy is the decision about whether a particular component of the product should be made in house or bought in. Largely determined by the capabilities and criteria established within the strategy, this decision becomes a routine matter that can be handled by a project team whenever it arises. The team should meet regularly to deal with this question, represent different functional views within the business and be familiar with the processes involved.

## 2.5 The traditional make or buy decision tree

Companies that have given serious thought to structuring their approach to make or buy can usually produce some kind of decision tree to guide project teams. This represents the trade-offs or choices that have to be made in considering the sourcing of a new product part or component. The decision tree is aimed at supporting the tactical and component category decisions that are made routinely at the operational level in the organisation. A typical example is shown in Fig. 2.3.

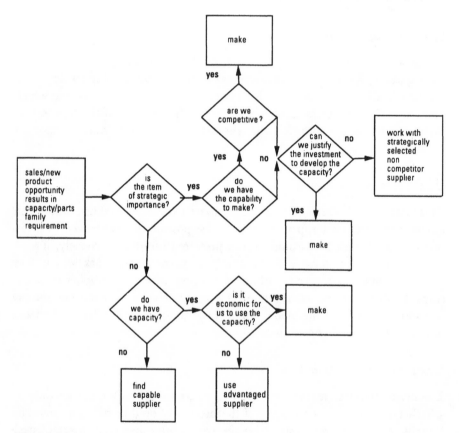

*Fig. 2.3 Make or buy decision tree*

The important thing to notice about such decision trees is that they nearly always assume that the strategic decisions have already been made, i.e. that the company and its employees have a clear idea of what it is important to keep in house. In practice this is rarely the case, and the project team using the tree finds that it is having to grapple with the strategic decision every time it works through the route. It is in precisely this area that the approach to formulating make or buy strategy which is described in this book can be useful. By having the strategy made explicit, the tactical and component decisions can be actioned by the type of decision tree shown in Fig. 2.3

## 2.6 What factors will need to be taken into account in a make or buy review?

The multidisciplinary nature of make or buy issues has already been mentioned. This is a reflection of the many different factors that need to be considered in developing a new make or buy strategy. Some of the principle factors are:

- market position and trends;
- company product and process capability;
- customers, competitors and suppliers — their characteristics, requirements and capabilities;
- cost analysis and comparison with the outside world;
- projection of financial results and sensitivity analysis.

The process of evaluation which brings together the consideration of these many factors in a manufacturing business is best carried out by a team of people representing the various functions or activities within the company. This process will be described in more detail in the following chapters, but first we need to investigate the content of the various factors. There are many fields of practical and academic work which have a contribution to make to the ideas. Individuals working in the manufacturing, finance, marketing, purchasing, and other functions in a company may not be familiar with all of these sources of ideas and their content, so an initial survey is useful.

## 2.7 Formulating a make or buy strategy in a manufacturing business

The concept of manufacturing strategy and its associated decision areas have been discussed earlier. The relevance for the make or buy process, apart from the realisation of its central importance, is the need to take into account the impact on all the other decision areas. The structural nature of make or buy means that nearly all other areas of business and manufacturing decision making will be affected. The introduction of new products is one of the most important, but also the size and location of plant and equipment and the consequence for investment decisions are others.

Within make or buy we are also including the consideration of customer/ supplier relationships. This is an area where there has been significant interest in recent years, and much research effort has gone into trying to understand the factors that make for successful relationships. The example of Japanese supply networks is often quoted to show how companies can work together to develop and make products very effectively, but without ownership stakes. Openness, mutual problem solving and trust are all said to be the characteristics of these ideal relationships from which business benefits flow. In the UK there have been a number of initiatives aimed at helping companies to develop this approach for themselves. For example, the CBI Partnership Sourcing programme and the work of the Glasgow Supply Chain Group [7].

Research into the origin and nature of these relationships in Japanese industry [8] indicates that there is no quick route to arriving at a strong and productive relationship. In many cases the industry form that supports this network of relationships has developed over decades rather than years, and there may be ownership stakes behind the scenes in the financial structure (keiretsu) of the industry which are not immediately apparent. Further, the incentive for continual cost reduction may have something to do with the balance of power in the relationship, rather than an idealised sharing of benefits.

A very important aspect in successful risk-sharing customer/supplier relationships is the element of trust. Research has also shown that trust builds up over time as experience is accumulated, and that at least three different categories of trust are required for the sort of robust relationship which will endure through the difficulties that will almost certainly arise. These three levels develop sequentially and are essential to enduring success. They are: competence trust, contractual trust and goodwill trust.

*Competence trust*

This the trust that each partner can actually do what they say they can do, i.e. that they can provide the service to the specified level of capability. Usually this is a matter of visiting the site, investigating manufacturing capability, running trials and finally monitoring performance. The aim, of course, is that when the relationship reaches a sufficient level of trust, then such monitoring becomes unnecessary and costs are saved.

*Contractual trust*

Trust that the supplier will stick to the terms of the contract through thick and thin, and not look for ways out of it if difficulties arise, is contractual trust. This trust takes a little longer to build up, because the experience of different contract conditions over a period of time is necessary to show that it really exists. Again, once developed this trust leads to cost reduction because less effort has to go into setting up complex contracts which foresee every possibility for difficulty.

## Goodwill trust

Goodwill trust results from the previous two forms of trust over a period of time, and means that both partners are prepared to give something to resolving the problems which inevitably arise in any long term business transaction, and will not manipulate the situation for individual advantage. Once achieved, this level of trust has major benefits for cost reduction and continued problem solving.

These forms of supplier relationship are only one possible outcome of the make or buy review. It may be decided that we wish to be highly vertically integrated, with all major processes within our own business. In fact the results of the make or buy review can take a whole variety of forms, as illustrated in Fig. 2.4.

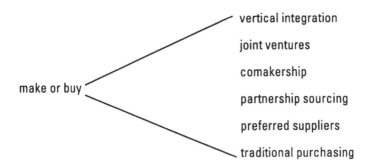

*Fig. 2.4 The sourcing options resulting from a make or buy review*

Depending on which part of this spectrum we are working in, and that will depend in turn on the importance of what is being sourced and whether we have decided to go outside the company at all, the type of customer/supplier relationship will vary considerably. Much recent work has gone into understanding the intercompany relationship aspects of the organisational forms at the centre of the spectrum. Traditional purchasing is the area concerned with buying commodities at the cheapest available price, and will not be developed further here. However, the issues involved in the consideration of vertical integration have been explored from many perspectives, notably industrial economics and strategic management.

## 2.8 Industrial economics

The major contribution from industrial economics has been in the ideas of transaction costs. Under this theory, developed by Oliver Williamson and expounded in his book 'Markets and hierarchies: analysis and antitrust implications'[9], and later publications, the determining factor in whether an activity is carried out inside or outside the firm is the cost of arranging the contract for the activity to be carried out. These are the transaction costs. In principle the hypothesis is that the mode of transaction which generates the least costs will determine the most successful organisational form. Williamson further suggests that transaction difficulties and hence costs arise from the following factors:

### Bounded rationality

The human mind is limited in its ability to deal with the complexity of real situations.

### Opportunism

Individuals and groups will seek personal advantage outside the terms of any agreement if the chance arises.

### Uncertainty

The full range of contingencies cannot be foreseen.

### Small numbers

Many bargaining situations are infrequent or involve small quantities where the cost of obtaining full information is prohibitive.

### Information impactedness

The information necessary to form a complete contract may be inaccessible in another organisation.

In order to limit the effects of these sources of transaction costs the firm will integrate and bring the provision of goods or services under its own control. However, there will be limits to the benefits of this process as Williamson himself states: 'The distinctive powers of internal organisation are impaired and transactional diseconomies are incurred as firm size and the degree of vertical integration are progressively extended, organization form held constant' (Reference 9, p.117).

Clearly there is an implication here that with growth and increasing vertical integration, changes in organisational form will be necessary to overcome the increasing internal costs of operation within the firm.

The application of the ideas of transaction cost economics to practical decision making remains limited, although some work has been done to operationalise them, for example Monteverde and Teece[10]. However, the important factor as far as make or buy strategy is concerned is the identification of costs other than production costs which need to be evaluated when making comparisons inside and outside the firm. These costs may take the form of total acquisition costs in the manufacturing context, and their components will be examined in Chapter 3.

This is a useful concept in assessing the total cost of either making in house or buying from a supplier, although the apparent assumption of transaction cost economics that the production costs will be the same wherever manufacturing is carried out is not a good model of reality.

## 2.9 Business and competitive strategy

Vertical integration and the factors affecting the success of firms operating at different degrees of integration have been much studied and reported. The degree of vertical integration with which a company is operating can be quantified as the value added by the company as a percentage of total sales.

Industry wide studies have been carried out to examine the link between the level of vertical integration at which a firm operates and its profitability. The conclusion from the PIMS (profit impact of marketing strategy) database[11] is that return on investment is higher at high and low levels of vertical integration, and that the middle ground (around 50% value added as a percentage of sales) is less profitable. This study was based on business units in western companies in the 1970s and 1980s, and hence may not be the best guide to current success, but there is an echo of this phenomenon in the performance of many current manufacturing organisations. While being to some extent industry sector dependent, companies with a low level of vertical integration are able to concentrate on those activities at which they excel or which are critical to their success, and thus be more profitable, and companies which are highly vertically integrated may be capturing more profit opportunities along the value chain. In contrast, companies in the middle ground may have no very clear strategy and be unable to maximise their profit potential.

Fig. 2.5 indicates the typical levels of vertical integration for market leading companies in particular industry sectors. Companies seeking to understand their level of vertical integration and make comparisons with

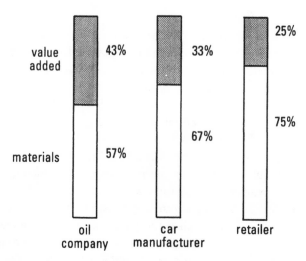

*Fig. 2.5 Typical levels of vertical integration vary between sectors*

leading competitors should be careful to do so within the bounds of comparable activities. It makes no practical sense for a retailer to try to emulate the level of value added achieved by an oil company!

The idea of the value chain was developed by, among others, Porter [12], to show where value is added during the progress from raw materials to finished goods. It is a useful concept, particularly in the manufacturing context, because by identifying the high value adding processes we can reveal which areas to protect from competitors, or to attempt to reduce cost and hence improve profit. The analysis can be carried out within a manufacturing company to examine the stages in the manufacturing process itself, or it may be applied to the whole industry of which the company is part in order to identify profit opportunities outside the current boundaries of the firm. Although the analysis provides good insights when carried out, it is not easy to do. The usual problem is that the costing systems in use in the company are not geared to allocating value at the various stages of production, and it may be difficult to establish values which can be meaningfully compared with the outside world. Fig. 2.6 shows an example value chain analysis for bicycle manufacture, with the build up of the value of the product as it moves through the factory. Each manufacturing process adds some value to the developing bicycle, until it is finally in a form in which it is sold to the outside world, in this case a wholesaler. The true value of the developing product at any point in the overall process is determined by what price it would command on the open market. In practice part-finished product is rarely sold, and so for internal accounting purposes the value is calculated using the factory costing system.

This determines the cost of each manufacturing process, including the overheads allocated to it, and makes the assumption that the cost incurred is equivalent to the increase in value.

Industry wide studies of the factors affecting the optimal level of vertical integration have also been carried out by Harrigan in the early 1980s[13]. The factors that she identified as significant were very much at the macro level in terms of industry conditions rather than in-company characteristics. This may provide some overall guidance to management about when it could be advantageous to increase or decrease the level of vertical integration, but it does not assist very much with the specific decision about individual products or processes which could be moved inside or outside the firm. The factors considered were: phase of industry development, volatility of industry structure, bargaining power and strategy objectives (leadership or focus). Perhaps the important message from this research for the make or buy strategist is that the slavish emulation of a competitor's level of vertical integration may not produce a comparable level of business success. The particular strategic objectives of the individual firm are also an important factor.

## 2.10 Strategic management

In recent years the strategic management literature has developed the ideas of competence within an organisation, and companies have given considerable effort to trying to identify their core competences. An example of this view of strategic management is contained in the article 'The core competence of the corporation' published in the *Harvard Business Review* in 1990, and written by C.K. Prahalad and Gary Hamel[14]. Competences are the particular capabilities individual to the firm, on which its ability to compete is based. Because competences may be combinations of skills, technologies, resources and ways of working, they are difficult to identify unambiguously, and even more difficult to manage.

The task of operationalising the concept of competences in the strategic management of a business is not a simple one. However, one of the more practical findings from this area of work is that at a business unit level it is the technologies that a firm can employ which are critical and that the scope to change these is limited. The identification of the technologies that a company is using is more straightforward than competence definition, although not without its difficulties. Companies usually find it easier to work with a hard definition of technology which includes materials, manufacturing processes, information and component technology, i.e. technology which is embedded within the product. Wider definitions of technology which include

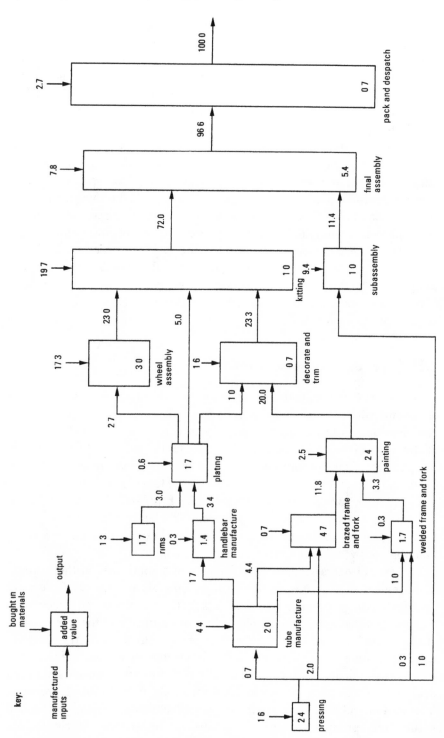

*Fig. 2.6 Example manufacturing value chain—bicycles*

organisational and decision making aspects begin to encounter the same difficulties as competence analysis.

Analysis of the manufacturing technologies (or manufacturing processes) can be a very powerful tool in the formulation of make or buy strategy. This is because their development and implementation is a matter of long time scale and often considerable investment, and they have a long term determining effect on the manufacturing capability of the firm. An analytical technique has been suggested by Pier Abetti, whose 'Linking technology and business strategy'[15] was published in 1989. This is a general discussion of how technological considerations can be incorporated into strategy formulation, based on both an academic perspective and Abetti's own personal experience of many years in the General Electric company in the USA. The technique considers two main aspects of the technologies in use by the company. First, the competitiveness with which the technology is deployed. This involves assessing the comparative level of performance that the company is achieving in the use of the technology. Investigation of other users and current best practice is necessary. Secondly, the importance of the technology to the business of the company is assessed. This requires a means of determining the impact of the technology on the factors that are central to the firm's success in the market place.

The technique involves displaying this information on a matrix with axes for competitiveness and importance (see Fig.2.7).

The scaling of the axes so that technologies can be accurately positioned is an area where we have developed techniques to support the project team. It is very important that the positioning of technologies on this matrix can be defended, as this is the focus of much debate about where the company should be investing its effort. A supporting methodology for scaling these axes enables arguments to be conducted on the basis of factual information, and for conclusions to be revisited in case of disagreement or new information arising.

## 2.11 A structured approach to make or buy strategy

The topics covered in this chapter will have given some idea of the range of issues to be dealt with in a strategic review of make or buy. Given the great variety of subject matter and experience that needs to be incorporated in the review, the need for a structured approach becomes apparent. A process containing a series of steps which successively introduce the various factors for consideration, enabling all the different interest groups in a company to have their say, is the goal we are aiming for. At the same time we need to include extensive investigation of market conditions, the capabilities of

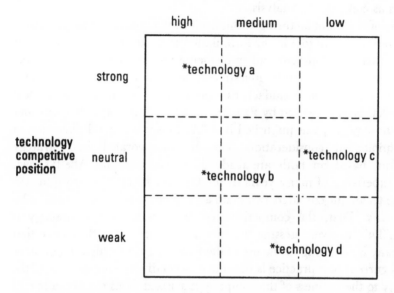

*Fig. 2.7 The technology competitiveness/importance matrix*

suppliers and competitors, the requirements of customers, and an impartial assessment of our own level of performance, strengths and weaknesses. This impartiality is difficult to achieve, since we are usually conditioned by the norms and beliefs of our own organisation. Bringing an external viewpoint to the assessment of our own company is a valuable additional contribution. This can be done through the involvement of independent people such as customers, colleagues from different parts of the company or external consultants. The nature and content of such a process will be discussed in Chapter 3.

*Chapter 3*

# Make or buy — the factors in a strategic review

## 3.1 Strategic review

There are four main phases to the strategic review of manufacturing operations that is the basis of the formulation of a new make or buy strategy:

phase 1: initial business appraisal
phase 2: internal and external analysis
phase 3: generation and evaluation of strategic options
phase 4: choosing the optimal strategy

The review is a team based activity, with the project team reporting directly to senior management. Individuals are selected to take part in the team on a combination of personal skills and knowledge of the business. More details of the team establishment and project setup are given in Chapter 5.

The purpose and content of each phase is shown in Fig.3.1.

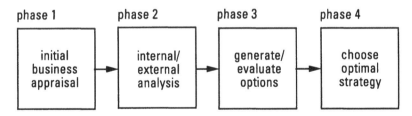

*Fig. 3.1 Phases of the strategic review*

## Phase 1: initial business appraisal

This is an assessment or, if recent work has been done in this area, a confirmation of the firm's overall business purpose and direction. This involves asking some fundamental questions such as:

what business are we in?
what do customers really want from us, why do they buy?
what are the trends affecting our markets?
where do we want our business to go?
what is our vision of the future?

In some companies these or similar questions are regularly discussed, and a clear direction is already set. However, in most cases the answers to such questions will not be widely known, even if senior management believes that it has a clear vision. In these cases the review offers an opportunity to clarify these issues, and more importantly to communicate the direction throughout the company.

Reaching sound conclusions to such questions requires more than a short discussion or statements of opinion from the managing director. In particular, the views of customers need to be carefully researched, and this may continue throughout the review. The important thing is to start with what is known to give an intial direction, and then to test and build on this during the further course of the project. There may be new insights which emerge from the review as it progresses which challenge or alter the initial perception. A willingness to accept new input and shift position is a fundamental requirement of the review process. This attitude must be set from the top of the company and made clear to all those involved.

Phase 1 requires more direct senior management involvement than any other in the review. Depending on how clearly management feels that it has already resolved the issues, it may choose different approaches to their elucidation and communication. A tentative initial opinion and direction offered to the project team with the brief to test and evaluate it, and then report back, is much more useful than a dogmatic expression of belief. However, if similar work has recently be carried out in the business, a sound basis may already be in place.

## Phase 2: internal and external analysis

This is the heart of the review in the sense that it is the engine which produces all the information on which later decisions are taken. Depending on the complexity of the industry structure, considerable effort may be involved. Details of the company's level of performance compared with competitors will be required. This is not only in terms of the price and quality of goods and services produced, but more difficult, also in terms of the level of performance achieved with internal processes. This means the assessment of manufacturing processes and other activities which are carried out to produce and deliver the goods but which the customer never sees.

Similar information is required of current and potential suppliers, so that comparisons can be made before sourcing decisions are finalised.

### Phase 3: generation and evaluation of strategic options

Having got a clear view of what the customer wants, the comparative performance of systems and processes and how these contribute to satisfying the customer, we are in a position to make some assessment of the possibilities open for change. This will involve considering the make-in and buyout options for each manufacturing process, parts family and subsystem in the product. The impact on the factors important to customers is assessed, and also the effect on measures of business performance that are significant to the company. Overall conclusions cannot be reached until the interactions of individual sourcing options are evaluated.

### Phase 4: choosing optimal strategy

During this phase different combinations of individual options are considered in order to establish the optimal new strategy. Financial models of the business are used to project the consequences of the various combined options. Although implementation planning is considered as part of the evaluation, the seeds of successful implementation are sown much earlier in the review. The involvement of key individuals from around the business, and the appropriate dissemination of emerging options during the project, will greatly enhance the likelihood of effective translation of the intended strategy into reality.

Within these phases and during the whole review there are a number of key aspects which are worth exploring, since they involve particular techniques or understanding.

## 3.2 Understanding business direction and objectives

It can never be assumed that the business has a clear direction and objectives. Even less likely is that everyone in the business will know what they are, even if they exist. Right at the start of the review process, it is essential to get this established. It may take some time and could continue throughout the project, particularly in terms of spreading the message and getting everyone involved. There is no substitute for senior management being involved with this process, since it is their direct responsibility to give this vision.

In its simplest form, establishing direction may just require the management team of the business to have a discussion together and reach concensus on the basis of their shared knowledge and experience. Unfortunately this often proves to be an unrealistic goal, since vested interests may impede agreement or, more likely, because of the sheer difficulty of the task. In these cases some facilitation of the process is required, in the form of supporting activities or moderation of the discussion. The activities are usually analytical techniques which can be applied by the management team to help gather information, clarify the discussion and reconcile opposing views.

The range and application of these techniques is described in a little more detail in Chapter 4, but they include the following:

## *Business scope and position*

An exploration of what our business does for the customer and our position in the industry value chain.

## *Key success factors*

Linked to the manufacturing objectives, these are the parameters that guide customer choice.

## *Business strategic issues*

Usually derived from a SWOT (strengths, weaknesses, opportunities, threats) analysis, these are the main issues of concern facing the business in its competitive environment.

## *Business activities analysis*

This is an assessment of the contribution to the business of the various activities that form the overall business processes. For example, this could include marketing, design and development, purchasing, manufacturing, sales and distribution, and after sales support. It may even be useful to carry out this first assessment before embarking on the make or buy project at all, because it may emerge, for example, that the business is all about selling and support, with little scope to influence success via manufacturing. This is an unlikely outcome, but the principle of establishing the main areas of importance for the business before investing a lot of effort in secondary activities is a good one.

## 3.3 Understanding the basis of manufacturing competitiveness: the manufacturing technologies

The review of literature and practice in Chapter 2 indicated the crucial role of manufacturing technologies in establishing a manufacturing capability. Assessment of the current and future contribution of the technologies to business success is a central theme in developing a new make or buy strategy. Visualisation of this analysis on the competitiveness/importance matrix is a very powerful way of prioritising technology sourcing decisions. The matrix itself suggests some generic sourcing strategies for technologies, depending on where they are found on the matrix, as shown in Fig. 3.3.

importance to business

| competitive position | | high | medium | low |
|---|---|---|---|---|
| | strong | invest maintain | consolidate keep pace | capability may open new market opportunities |
| | neutral | invest develop | partnership | stop outsource |
| | weak | initiate R&D or cease, find comaker | partnership | stop sell/licence design out |

*Fig. 3.2 Generic sourcing strategies*

For example, technologies in the top left hand corner of the matrix are vital to business success, and are likely to embody the company's core knowledge and experience. These are the technologies that the company should continue to invest in, and protect at all costs from competitors and suppliers. Bottom right technologies are probably anachronisms, left over from earlier product generations when they had some value to the business. The question now is how to get rid of them, they are just a distraction to the business. It may be that some value can be realised for them, although this is not the main priority.

Bottom left technologies are a more serious problem. They represent areas of great importance to the business, but where performance is weak. Each case must be examined individually for investment to bring capability up to best practice levels. Alternatively, if we cannot afford to develop the capability in house, a very close relationship with a technology provider should be sought. High levels of trust will be the main characteristic of this relationship, since we are dealing with technology so central to our business fortunes. Comakership relationships of this kind require very careful selection of partners.

Top right technologies may be a source of great opportunity. Clearly, this is an area where we have an unexploited asset. What we do with it depends on the possibilities open to the business of realising some value from the asset. Although is may be tempting to branch out into new products, markets or even new businesses, these options should be carefully considered in terms of the distraction they may cause from the real focus of the business already identified. It may well be better to license the technology out or form a separate joint venture, so that this distraction is minimised.

Central areas of the matrix are the obvious areas for partnering with suppliers. This partnership could take a variety of forms, depending on the particular issues of the technology concerned. The various forms of partnership and supplier relationship willl be discussed in the following section.

The main benefit of the matrix is that it introduces some prioritisation for which technologies should be outsourced and which retained in house. The final choice will be dependent on a number of other factors including the interaction between technologies and the finance available. These will be explored in subsequent chapters, but for now the important aspect is the first prioritisation introduced by the competitiveness/importance matrix. The assessment of the scoring of the axes of the matrix requires a degree of rigour in the analysis which is developed in Chapter 4, and which occupies a considerable amount of project time.

In addition to the assessment of manufacturing technologies, it is essential to have a clear understanding of the product structure — the product architecture.

## 3.4 Understanding the nature of the product: the product architecture

The company may make one or a number of product families, each characterised by its individual product architecture. The architecture is the generic structure of the product, in terms of the subsystems and parts families of which it is composed. In this breakdown of the product structure we are not concerned with part number or drawing level detail, but rather the functional description of the main components of the product. The purpose is to identify the functional subunits of the products in a way that they can be considered for sourcing choices. The need to do this arises from the fact that although technology sourcing decisions are vital for long term strategic reasons, it is very often whole subsystems and/or parts families which are outsourced, rather than discrete manufacturing technologies. There will be a strong influence of one on the other, as outsourcing a complete parts family may render a manufacturing technology superfluous in house. What we need to do is consider technology, subsystem and parts family sourcing as an integrated whole, with the interactions between them identified.

The relationship between the technologies, subsytems and parts families may be imagined as a cube, with these as the axes, as shown in Fig. 3.3.

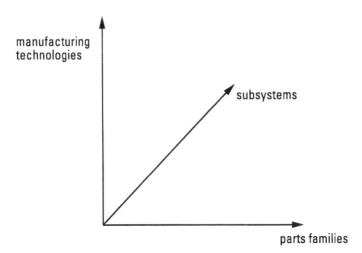

*Fig. 3.3 Manufacturing technologies, product subsystems and parts families considered as the axes of a matrix*

The competitiveness/importance matrix can be applied to the assessment of subsystems and parts families in a similar way as to technologies, to provide a first indication of the sourcing priority of the various product components. The scoring of the axes is the same process as for the technologies.

The initial process of breaking down the product architecture is best guided by taking a functional view of the product. An example for the family car is shown in Fig. 3.4.

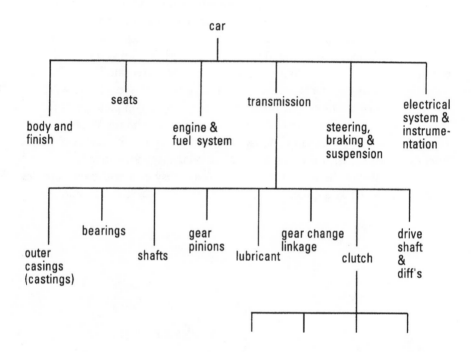

*Fig. 3.4 Example product architecture: family car*

The breakdown of the product structure can continue until individual parts or parts family level is reached. This will allow an assessment of the commonality of parts content of the whole product to be made, and a link to the manufacturing technologies to be established. For instance, within the clutch there are bearings, pressed metal parts and turned parts.

A useful technique which can be applied to assist in the development of the product architecture is quality function deployment (QFD), a more detailed description of which is given in Chapter 7. QFD can be used to assess critical design characteristics against customer requirements, and produce a ranking. These rankings can be used in the assessment of the subsystem and parts family characteristics. Carried out at a top level this analysis can help to validate the choice of subsystems and parts families shown in the product architecture tree, through identifying which of these has most impact on meeting customer requirements.

## 3.5 Suppliers and supplier relationships

The area in the centre of the competitiveness/importance matrix has been identified as the region where it is particularly important to explore a variety of supplier relationships. Depending on the degree of cooperation and trust necessary, there is a whole spectrum of possible relationships, as shown earlier in Fig. 2.4.

These options represent increasing levels of involvement, moving up the list. At the top, vertical integration and joint ventures both involve ownership of the supplier, and will not the considered further here. At the other extreme, traditional purchasing tends to be an adversarial relationship, where price is the single most important factor in selecting a supplier. Little trust or loyalty is involved, and on completion of a contract source of supply may be abruptly switched.

The area in the middle is where much effort has been expended in recent years, as companies work more closely with selected suppliers. The purpose is to achieve some of the performance and control of supply benefits which would formerly have been associated with ownership, but without the commitment of capital or the risk of asset exposure.

Comakership is a very close and committed relationship characterised by openness and shared risk. There will be open book accounting between the two companies, as they cooperate in the development of new products and the continuous improvement of existing products.

Partnership suppliers are involved in the product development process, but cost transparency is not at the level of comakership.

Preferred suppliers have been actively selected for capabilities beyond best price, and a long term working relationship is anticipated. There will be an expectation of mutual problem solving and ongoing cost reduction of the product. The supplier will be able to meet consistent quality standards, reducing the need for goods inwards inspection and checking at the receiver. In moving from a traditional purchasing policy to one of working with preferred suppliers, many companies have reduced their number of suppliers by factors of five to ten.

Recent survey work has indicated that traditional purchasing is now the minority form of relationship with suppliers among large UK manufacturing companies. The move to develop closer working relationships with fewer suppliers goes hand in hand with a strategic approach to make or buy. However, the important factor to bear in mind is that a make or buy strategy which determines the manufacturing capability of a company needs to be taking a view over many months or even years, and supplier relationships should be considered over the same timescale.

# 3.6 Costs

This aspect of make or buy analysis is the one which usually receives more than its fair share of attention. Traditional approaches to make or buy concentrate on cost comparisons between making a product in house or buying it from suppliers. Much has been written on this subject, particularly from an accounting perspective. The calculation of varying financial contributions when making the choice of which process or item to outsource due to capacity constraints is a well developed area.

The difficulty with this approach is that, although the costs are easy to compare, the basis of calculation may be quite different. As a result, there is often an issue of whether we are comparing like with like. There are a number of cost categories which need to be considered, including in house manufacturing costs, acquisition costs and suppliers quoted costs. We shall consider each briefly here.

## In house manufacturing costs

The big problem with costing systems which are already in place is usually the arbitrary allocation of overheads or other activity costs from within the organisation. If we are trying to establish our costs for a particular manufacturing process this can lead to huge variation compared with the outside world. The scope to influence the final apparent cost of the process is illustrated by the example cost build up shown in Fig. 3.5.

Costs which cannot easily be directly linked to the process are usually allocated as overheads. The method by which this allocation is made is the source of the variation; it may or may not fairly reflect the process's utilisation of other resources in the company.

During the project to develop a new make or buy strategy for a manufacturing business, particular attention should be given to reviewing the cost structure. In most cases it is useful to develop a new cost model of the manufacturing processes from scratch, in order to be certain that comparisons are well founded. The detail of this will be described in Chapter 4.

## Acquisition costs

In making a true assessment of bought in costs, it is not enough to consider just the supplier's quoted price. In order to compare fairly with internal cost, we need to take account of the costs of bringing the purchased item or service into the organisation: the acquisition costs. Often overlooked, these costs can add between 10% and 30% to the quoted price. The source of these costs

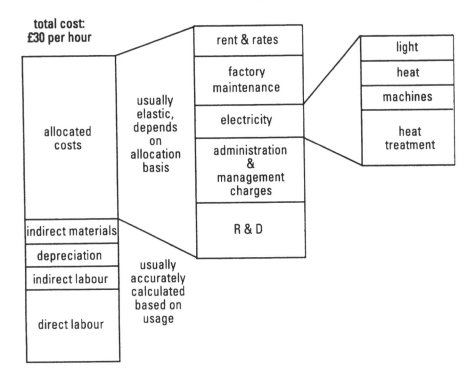

*Fig. 3.5 Example of a cost build up for a manufacturing process*

includes the buying office, drawing up contracts, legal costs, delivery, goods in, storage and inspection.

It is good practice to include a calculation of acquisition costs in the new manufacturing cost model, so that more complete comparisons can be made with suppliers' quotes.

## Suppliers' costs

It is not sufficient to assess suppliers' quotes simply on price and acquisition costs. Before making a comparison with in house costs we need to know more about quality and delivery performance. Is the supplier using the same manufacturing process, what level of quality approval does it have, what lead time and stock holding can it offer? Over what period of time is the company prepared to guarantee this level of performance? All these factors need to be thoroughly investigated before we can be confident of the basis on which cost comparisons are being made.

## Volume effect on costs

It can easily be overlooked that the calculation of cost levels is heavily dependent on the level of utilisation of the factory. At high and low levels of utilisation the volume/cost sensitivity is particularly acute, and great care should be taken in coming to any conclusions here. Incremental cost, i.e. how much it costs to produce one more unit of output, is the critical measure. For a certain level of fixed costs, i.e. a certain installed capacity, the incremental cost is steady for the normal range of plant utilisation, say between about 60% to 80% of theoretical maximum capacity. However at very low levels of utilisation (0% to 20%) and very high levels of utilisation (80% to 100%) the incremental cost rises steeply. At these extremes make-in/buyout decisions are likely to be poorly based if volume/cost sensitivity is not taken fully into account. Stable make or buy decisions can only be made in the flat area of the incremental cost curve shown in Fig. 3.6.

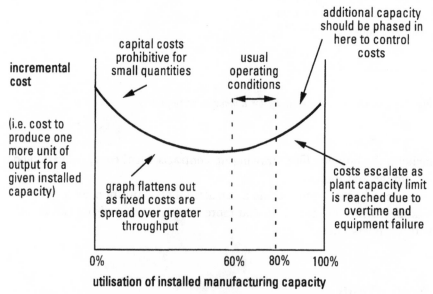

*Fig. 3.6 Effect of capacity utilisation on incremental cost*

# 3.7 Analytical techniques in the development of a make or buy strategy

The project work involved in developing a new make or buy strategy can involve a considerable amount of analytical detail. In order not to become unnecessarily trapped in this, every opportunity should be taken to simplify the task, selecting only those areas for detailed analysis which are critical to the result. The Pareto principle is particularly useful; i.e. analysis of 20% of the data may lead you 80% to the conclusion, and this should be applied whenever possible.

A number of analytical techniques which are similarly useful in the strategy development process are described in Chapter 7. However, the key requirement is not to get bogged down in the limitless possibilities for analysis, but to remain focused on the main purpose of the project. Some of the practical ways to do this are as follows:

(i) Make hypotheses about the likely outcomes of the project and focus the analysis to test and validate these hypotheses.

(ii) Carry out sensitivity checks as additional analysis is made: i.e. does taking a particular analytical task further forward actually make much difference to the emerging conclusion?

(iii) Use some form of risk assessment to determine which areas of analsyis are critical to a soundly based outcome from the project. This can take a quite simple form such as 'what if?' scenario planning, i.e. imagining the worst case circumstances which the new make or buy strategy might have to work within. This can be thought of as setting some boundary conditions on the likely future business environment.

## 3.8 Maintaining the strategy

Having devised a strategy that is optimal for the business goals, it is important to keep it up to date and abreast of a changing environment. This is best done by identifying the critical conditions which have shaped the strategy, and then monitoring them for any changes. The likely most important are likely to be:

- a change in business strategy;
- internal load changes;
- major new design projects;
- new production equipment investment plans;
- supplier quality and delivery ratings;
- supplier capacity changes;
- supplier price changes.

By maintaining a watching brief on these factors, and a record of how the current strategy was devised, the company is in a good position to respond rapidly to changing conditions. The consequences for the business can be assessed, and if necesssary a revision of the strategy carried out. The details of the mechanisms for monitoring and responding to changes in conditions will be described in Chapter 5.

*Chapter 4*

# Developing a make or buy strategy — guide to a practical approach

## 4.1 Introduction

The factors described in Chapter 3 can all be integrated into a systematic review of the make or buy requirements for a manufacturing business. The main stages of initial business appraisal, internal and external analysis, the generation and evaluation of options and the choice of an optimal strategy, have all been outlined. In order to assist a project team to work through these systematically, a more detailed breakdown is required. The steps in this breakdown are shown in Fig. 4.1. Each step will be introduced with a flow diagram to show the principal activities within it, and then described in working detail, with its objective, the approach to follow, and hints and tips from practical experience. The steps are:

1. Derive business issues
2. Data collection
3. Develop ideal greenfield business scenario
4. Product architecture definition
5. Manufacturing technology definition
6. Technology cost modelling
7. Architecture/technology relationship
8. Manufacturing technology assessment
9. Product architecture assessment
10. Decision support modelling
11. Evaluate technology, subsystem and parts family options
12. Develop strategy recommendations and implications

The steps are not necessarily followed in a purely sequential manner; in many cases there will be iteration and parallel working between steps. Normally, however, by the end of the project the content of each step will have been covered, although depending on the precise business conditions some steps may be more important than others. For example in a business with relatively few manufacturing technologies, step 8, manufacturing

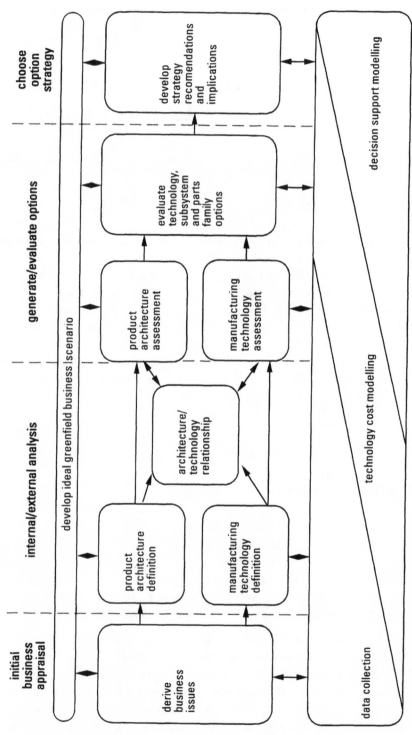

*Fig. 4.1 Strategic make versus buy methodology*

technology assessment, will require less work than one where there are many. A thorough understanding of the product architecture may, on the other hand, be critical, and step 9 would assume greater importance and correspondingly more effort.

## Step 1: derive business issues

### *Objective*

To establish the scope and strategic direction of the total business, involving the management team in defining the strategic issues facing the business.

### *Approach*

This first step in the methodology is critical to establishing the overall framework and direction for the project. It cannot be done without the full commitment of senior management as it draws on their knowledge of the business. Further, it requires their agreement to such fundamental conclusions as mission, objectives, markets, key success factors and strategic issues.

There is no one correct route to follow in reaching agreement here; the route chosen will depend on how well defined these aspects are already in the business concerned, and what degree of concensus exists among senior managers. In a situation where the management team believes that it has bottomed out these issues, there is just a need to run some independent checks on the quality, consistency and completeness of their conclusions. This can be done by reference to customers, external authorities (either corporate or independent) and competitors. In most cases there will be a need either to achieve concensus about existing views or to derive answers to these fundamental questions from scratch. There are many ways of tackling this, usually based on a workshop format. A summary of some of the issues to be addressed and some well proven tools and techniques is given here. The project team should select and use those techniques best suited to eliciting the missing information in their particular business, and the degree to which a team building process is required. A fuller guide to each technique is given in Chapter 7.

### *Issue: what is the business scope?*

This is best dealt with as a workshop which sets out to answer such questions as *what business are we in?*, *what are we really offering the customer?* and *how would we set up the business if we were starting again from scratch?*. A useful construct here

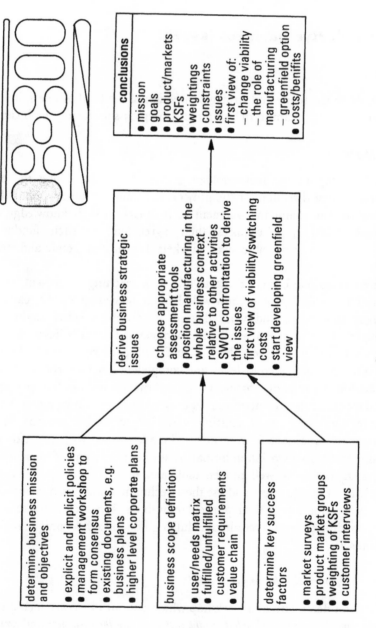

**determine business mission and objectives**

- explicit and implicit policies
- management workshop to form consensus
- existing documents, e.g. business plans
- higher level corporate plans

**business scope definition**

- user/needs matrix
- fulfilled/unfulfilled customer requirements
- value chain

**determine key success factors**

- market surveys
- product market groups
- weighting of KSFs
- customer interviews

**derive business strategic issues**

- choose appropriate assessment tools
- position manufacturing in the whole business context relative to other activities
- SWOT confrontation to derive the issues
- first view of viability/switching costs
- start developing greenfield view

**conclusions**

- mission
- goals
- product/markets
- KSFs
- weightings
- constraints
- issues
- first view of:
  - change viability
  - the role of manufacturing
  - greenfield option
- costs/benifits

*Step 1 Derive business issues*

is the users/needs matrix, which helps to define customer requirements and the degree to which our products or services satisfy them.

## *Issue: what are the key success factors in our business?*

Businesses fail or prosper in the market place according to how well they fulfill certain key success factors (KSFs). These are the parameters that guide customer choice, and the company which optimally meets these sometimes conflicting factors will win the lion's share of the market. The KSFs in manufacturing industry are normally variations on the theme of quality, cost, delivery and flexibility. They will vary between product/market groups for the same firm, and will have different weightings attached to each factor in a particular product/market group.

The identification of the KSFs for the product/market groups under review is an early requirement, together with their respective weightings.

The KSFs may in turn be broken down into elements and drivers, with associated output measures, to assist the process later in the methodology of evaluating how the firm's activities impact its achievement of the KSFs.

## *Issue: what are the business strategic issues that we have to deal with?*

A useful technique here is a rigorous SWOT (strengths, weaknessess, opportunities, threats) analysis to derive the principle strategic issues facing the business. It is carried out in a workshop, and is a powerful means of achieving concensus about the major issues facing the business. These issues can then be used later in the methodology as some of the criteria against which in and outsourcing options are assessed.

## *Technique: plotting the business activities matrix*

A necessary precursor to deciding how to concentrate manufacturing resources is to consider the position of manufacturing in the context of the whole business. For example we can consider the business to consist of six key activities:

marketing
design and development
acquisition and sourcing
manufacturing
sales and distribution
after sales support

These activities are assessed in terms of their links to the KSFs and depicted on an importance/competitiveness matrix. As a result we form an initial picture of whether the make versus buy strategy in manufacturing is really the area on which to concentrate effort, or whether there are more significant areas requiring management attention. A further important aspect to emerge from this initial analysis is the link between activities, for example between manufacturing and product design capability. This may be significant later when assessing the possibility of moving activities in or out of the business.

This analysis is a safety measure either to prevent unnecessary project work or focus required effort; it can also usefully be carried out at the diagnostic stage before the project is approved.

## Tool: DTI competitive manufacturing workbook

This Department of Trade and Industry book is the first stage in building a manufacturing strategy, and covers an audit of the existing strategy. It takes the form of a self assessment workbook, and the first six worksheets deal with the current business orientation, level of performance in manufacturing and scope for change.

## Technique: value chain analysis

It is essential to have a picture of how the company fits into the full industry sector. This can be shown with a high level value chain, and also expanded in detail for the firm itself. These charts are helpful in understanding what opportunities there may be to extend or restructure the range of operation of the firm, and also in defining what the goals of the firm are. Analytical effort should be concentrated on the part of the chain relevant to the company.

It is useful to compare the level of vertical integration that emerges from the value chain analysis of the firm with others in the same industry sector, or even other business units within the same company. This can give an early picture of how competitively structured the firm is at the time of comparison.

A useful additional output from the value chain analysis can be a first view of the likely viability of moves in or out of the business, i.e. a view of switching costs. There may be some activities or technologies where redundancy and equipment costs could never be recovered if they were to be outsourced.

## Hints and tips

- select the tool(s) and technique(s) to suit business and project conditions;

- thoroughly brief the management team on how the relevant process works;

- allow enough time for a free-ranging discussion and brainstorming;

- individuals must feel able to contribute in a non-judgmental atmosphere;

- high level facilitation skills are required;

- the full project team will not be necessary for some exercises.

## Step 2: data collection

### *Objective*

To assemble, review and summarise all relevant business data. This information is compiled to enable an account to be given of the business within its economic and competitive environment.

### *Approach*

The make or buy project has extensive data requirements, from both inside and outside the firm. The first task is to brainstorm all required data and possible sources. A plan of the data collection activity can then be made. It is particularly important to allow enough time for external data gathering; information from suppliers and about competitors may have an extended lead time.

In view of these large scale data requirements, it can be useful to focus the activity by hypothesizing possible outcomes of the review. In other words, there is little point collecting data in support of unlikely outcomes.

Within most companies there will be a key document such as an annual business plan or equivalent strategic planning document. In summary, the scope of the information required is shown in Table 4.1; this should be taken as a minimum requirement.

Some additional explanation about data elements and their sources may be useful:

### *Management accounts*

The last financial year's complete accounts should be used. These will be

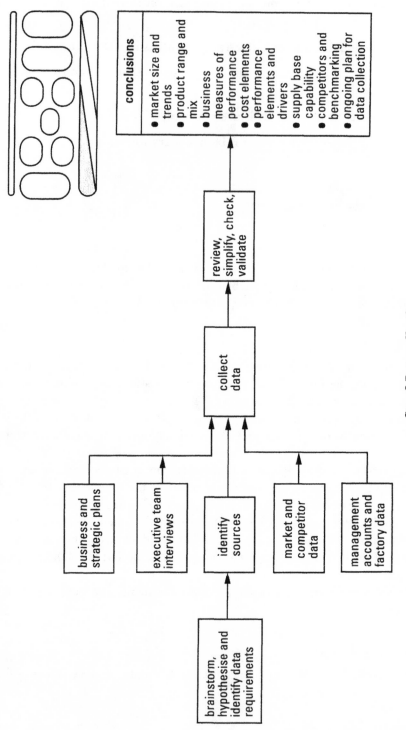

conclusions

- market size and trends
- product range and mix
- business measures of performance
- cost elements
- performance elements and drivers
- supply base capability
- competitors and benchmarking
- ongoing plan for data collection

review, simplify, check, validate

collect data

business and strategic plans

executive team interviews

identify sources

market and competitor data

management accounts and factory data

brainstorm, hypothesise and identify data requirements

*Step 2 Data collection*

*Table 4.1 Data collection*

| Data element | Comment | Source | Sanity check |
|---|---|---|---|
| **Market**<br>size<br>share<br>segments<br>competitors | Require current<br>and trend<br>information | Strategic plans<br>sales forecasts<br>customers | corporate view<br>customer view |
| **Product Groups**<br>description<br>size | check common<br>view between<br>manufacturing<br>and marketing | sales and<br>marketing,<br>planning and<br>control<br>documentation | general manager |
| **KSFs and OWC**<br>price<br>delivery<br>quality<br>features<br>specification etc. | yes/no; rank<br>*how do we*<br>*compare?*<br>*what are the*<br>*trends?* | customers<br>marketing<br>competitor<br>information | general manager<br>customers (end<br>users)<br>corporate<br>benchmarking<br>studies |
| **Business Measures of Performance (MOPs)**<br>ROCE<br>gross margin<br>sales/employee<br>AV/employee<br>stockturns<br>directs/indirects | Selected<br>according to how<br>the business is<br>being assessed | finance<br>general<br>manager<br>corporate | what measures<br>do competitors<br>use? |
| **Management Accounts**<br>labour costs<br>material costs<br>overheads etc. | see detail in step<br>6 | finance | finance |
| **Value Chain**<br>system value<br>subsystem value<br>assembly value<br>subassembly<br>value<br>component value | establishes a<br>picture of how<br>value is added<br>through the<br>business process | finance<br>manufacturing | customers<br>suppliers<br>corporate |

needed for consistency in the cost model and as a base from which to measure improvement.

### Supply base

The capability of current and potential suppliers. An understanding of the cost and performance drivers in the use of particular manufacturing technologies or the production of particular parts families and subsystems.

### KSFs and OWC

These are the key success factors described earlier, and the order winning criteria. The OWCs are useful in determining the aspects of product offering which really make a difference to customers.

### Competitors

An understanding of how competitors are performing against the key success factors, and any insights that can be gained into their cost base and methods of operation.

### Product and process

Functionality, technology trends, outside experts' views of the likely developments.

### Best practice

In addition to the capabilities of suppliers and competitors, there can be ideas to be picked up from nonrelated fields. Benchmarking studies of business practice and other industries' use of equivalent technologies can provide both insights into new practices and yardsticks to assess our competitive performance.

In addition to document sources, some of this information can be gathered from interviewing key personnel on the management team. This will also be helpful in ranking the key success factors and business measures of performance, i.e. are some more important than others?

A further important aspect of the data collection activity is validation. Wherever practical all data should be checked or validated by independent reference to another source. This could be by sample back to the origin, or cross checks with third parties in the case of qualitative data. Where validation is not possible, it is important to keep track of unchecked data in terms of their impact on strategy formulation. This is useful later in the sensitivity testing (steps 10, 11 and 12).

## *Hints and tips*

- involve the management team from an early stage;

- constantly review and evaluate gathered data for validity;

- validate base data with the management team before reaching any tentative conclusions;

- talk directly to customers to check internal perceptions;

- existing company plans cannot be assumed to be accurate or comprehensive.

## Step 3: develop ideal greenfield business scenario

### *Objective*

To identify the characteristics (size, location, capability) the manufacturing business would have if unconstrained by the inherited situation.

### *Approach*

The ideal greenfield view of capabilities (business activities, technologies, product range) provides a useful yardstick with which to assess the possible real world solutions that are subject to constraints. These constraints are, for example, location, previous investment, divestment restrictions, limited capital availability, social and political factors.

The greenfield scenario will develop through the project, and is shown this way in the step 3 diagram. An initial view should emerge from step 1, *derive business issues*. This will be based on the managment team's current knowledge of the profit opportunities in the industry's value chain together with the foreseen market and technology trends. In its first form it may be the result of a brainstorming session, or the generation of hypotheses. As the project progresses these initial views will be validated and modified. More detailed information on customers, competitors, suppliers and trends will be available. This enables the greenfield scenario to be updated regularly, until

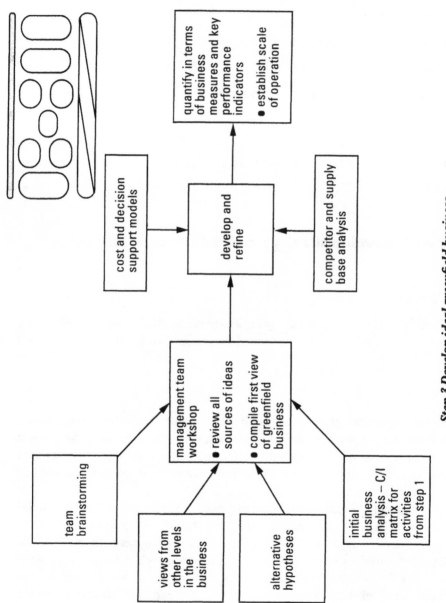

**Step 3 Develop ideal greenfield business**

it can be assessed by the criteria used in the decision support modelling (step 10). The evolving scenario will be regularly discussed at the management review meetings. The characteristics and financial consequences of the greenfield scenario may then be used as a measure of how optimal the final real world strategy options are.

## *Hints and tips*

- do not invest too much effort in detailing the first pass view, it will evolve; get views from all levels including the top management team and the taskforce team;

- keep the greenfield picture simple; it is intended to be a top level definition of the business, unlikely ever to be practical, and for easy comparison with real world solutions.

## Step 4: product architecture definition

### *Objective*

To derive the structure of the main products being manufactured by the business.

### *Approach*

Product architecture analysis is important in businesses where the product is complex, consisting at the next level of many subsystems and below that multiple parts families. For businesses operating in different product/market groups, there may be several generic product architectures to be analysed.

The make or buy review requires a thorough understanding of how the various parts of the product architecture contribute to satisfying customers and winning orders. There may be some subsystems and parts families which it is critical to retain in house and others that can be advantageously sourced from outside.

An initial step is to identify the generic products in each product/ market group served by the company. The product/market groups themselves should have been identified in step 1, and this now needs to be

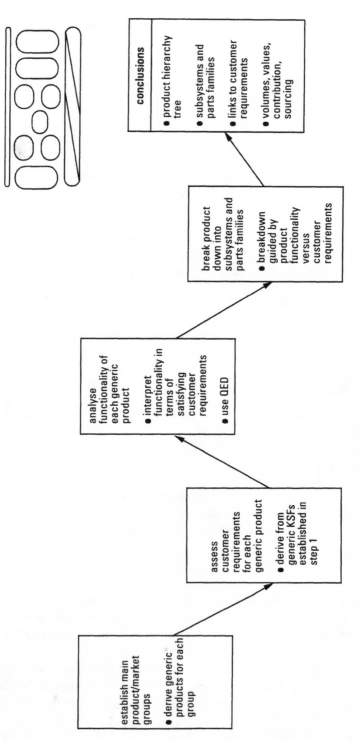

**establish main product/market groups**

● derive generic products for each group

**assess customer requirements for each generic product**

● derive from generic KSFs established in step 1

**analyse functionality of each generic product**

● interpret functionality in terms of satisfying customer requirements

● use QED

**break product down into subsystems and parts families**

● breakdown guided by product functionality versus customer requirements

**conclusions**

● product hierarchy tree

● subsystems and parts families

● links to customer requirements

● volumes, values, contribution, sourcing

*Step 4 Product architecture definition*

taken a little further to decide which are the generic products in each group. The customer requirements for each generic product will be closely linked to the key success factors for the business, also derived in step 1.

The approach is to consider the functionality of the product – i.e. what does it do for the customer? This basic question guides the subdivision into subsystems and parts families, resulting in a hierarchical tree which shows how the product is constituted. Fig.4.2 shows an example of the product architecture as it might be derived for a familiar consumer product, the family saloon car. In this example the transmission system has been further broken down and as a result some parts families have emerged, such as bearings and shafts. There are also some subsystems, such as the clutch and the gear change linkage, which could be broken down to identify additional parts families if this is useful in the later analysis.

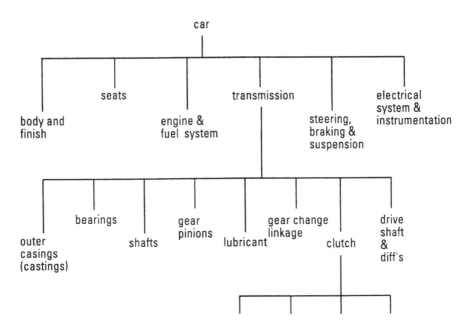

*Fig. 4.2 Example product architecture: family car*

Understanding of product functionality and its link to customer requirements may be facilitated by drawing top level input/output diagrams for the product, establishing critical design characteristics and by applying QFD (quality function deployment). An extension of the example of the family car to show how this analysis can be applied is given in Chapter 7.

## *Hints and tips*

- concentrate on main product market groups and generic products - use Pareto analysis to identify the significant volumes and groupings throughout;

- a workshop process can be used to help define and agree a workable product architecture;

- define and agree the list of subsystems and parts families;

- keep note of how subsystems and parts families link to manufacturing technologies when this becomes visible during the analysis - it will be needed later;

- parts families can be categorised in a number of ways: function, product linkage, standard hour load on process, shape, material, tolerance;

- choice of categorisation depends on the distinctive characteristics of the industry - but process links are always important.

# Step 5: manufacturing technology definition

## *Objective*

To define and list the manufacturing technologies used by the business. This includes technologies both internal and external to the business, and forms the basis of all subsequent technology analysis.

## *Approach*

The aim is to produce a working list which does not go into unnecessary detail, but which is adequate for the subsequent analysis. Since the amount of analysis is proportional to the number of technologies, there is a clear incentive to aggregrate where appropriate. A number of techniques are used to generate this list as follows:

- team brainstorm of internal and external technologies on the basis of existing knowledge;

- factory floor audit to identify technologies in use;

- interviews with manufacturing and design engineers;

- asset register review;

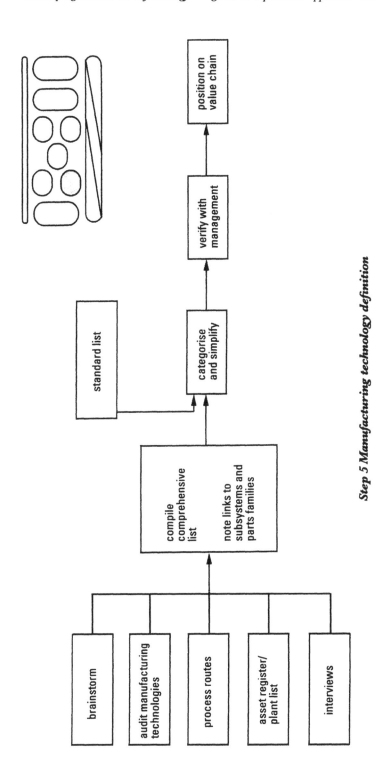

*Step 5 Manufacturing technology definition*

- process route analysis;

- standard manufacturing technology list (e.g. Reference 16, see Table 4.2).

Having assembled all possible technologies, the task is then to categorise and simplify them without losing any significant variations. For example, conventional machining may include drilling, tapping, honing, turning,

*Table 4.2 Classification of manufacturing technologies (adapted from Hayes and Wheelright 1984)*

Changing physical properties

| chemical reactions | refining/extraction | heat treatment |
| hot working | cold working | shot peening |

Changing the shape of materials

| casting | forging | extruding |
| rolling | drawing | squeezing |
| crushing | piercing | swaging |
| bending | shearing | spinning |
| stretch forming | roll forming | torch forming |
| explosive forming | electrohydraulic forming | magnetic forming |
| electroforming | powder metal forming | plastics molding |

Machining parts to a fixed dimension

traditional chip removal

| turning | planing | shaping |
| drilling | boring | reaming |
| swaging | broaching | milling |
| grinding | hobbing | routing |

Nontraditional machining

| ultrasonic | electrical discharge | electro-arc |
| optical lasers | electromechanical | chem-milling |
| abrasive jet cutting | electron beam | plasma-arc |

Obtaining a surface finish

| polishing | abrasive belt grinding | barrel tumbling |
| electroplating | honing | lapping |
| superfinishing | metal spraying | inorganic coatings |
| parkerising | anodising | sheradising |

Joining parts of materials

| welding | soldering | brazing |
| sintering | plugging | pressing |
| riveting | screw fastening | adhesive joining |

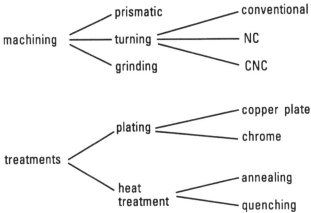

*Fig. 4.3 Defining manufacturing technologies*

value chain

| suppliers | company internal technologies | | | customers |
|---|---|---|---|---|

| | CNC turning<br>CNC milling<br>CNC grinding<br>conventional<br>machining<br>broaching | heat treatment<br>winding<br>electroplating<br>metal finishing | actuator<br>assembly<br><br>systems<br>assembly | systems<br>integration |

casting
forging
extruding/
drawing
metal forming
powder metal
forming
nonmetallic
moulding

| | EB welding<br>welding/brazing<br>gear manufacture<br>lapping | | painting<br><br>testing | painting<br><br>testing |

heat treatment
electric discharge
chemical machining
electrochem machining

sub contractors

CNC turning       fluidised bed      testing
CNC milling       heat
CNC grinding      treatment
conventional      electroplating
machining
broaching
gear manufacture

abrasive flow
electronic component manufacture

*Fig. 4.4 Positioning the manufacturing technologies on the value chain*

milling, grinding. The level at which manufacturing technologies should be defined will depend on local conditions. The aim should be to define at as high a level as possible to minimise the subsequent analysis, but without losing important distinctions. For example, see Fig.4.3.

Finally, the position of both internal and external technologies is plotted on the value chain for the business. This has a number of benefits: it gives a picture of the existing degree of vertical integration in manufacturing and shows which technologies are closest to the customer. By mapping the technologies against position in the flow of manufacturing we begin to gain some insight into the possible impact of changing technology sourcing, either in or out. Areas of technology duplication between ourselves and the outside world are also revealed, leading to rationalisation possibilities. An example of this plot is shown in Fig. 4.4.

## *Hints and tips*

- simplify the initial list of technologies compiled;
- it may be necessary to form secondary groupings:
  - high and low volume production
  - different materials being processed
  - each product group in a complex business;
- select the appropriate level of technology definition to simplify the analysis.

# Step 6: technology cost modelling

## *Objective*

To establish an accurate cost assessment of the manufacturing technologies used by the business, for comparison with outside suppliers.

Conventional costing systems used in factories are primarily concerned with recovering all the costs of running the business and allocating these to the products made. These are of little use in the make or buy decision since they do not generally reflect the true cost of manufacture; usually as a result of the arbitrary allocation of overheads.

We therefore require a manufacturing technology based cost model to

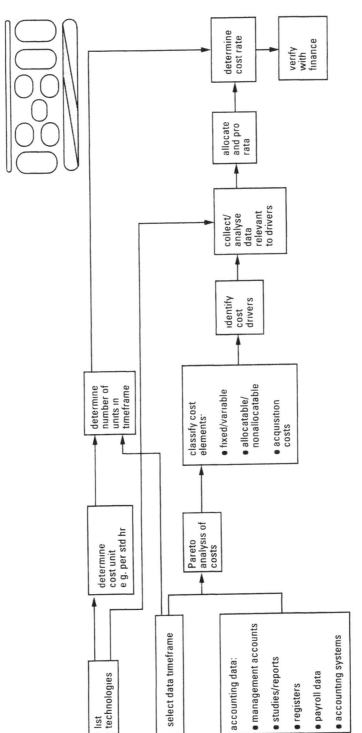

*Step 6 Technology cost modelling*

determine the real costs, addressing the realistic allocation of all business costs. The key output of this is a unit cost, e.g. cost per hour, for each technology. This cost figure is the basis for comparison with supplier rates.

An additional important output of this cost modelling for the purpose of the make or buy strategy is an assessment of total acquisition costs for the business under review.

## Approach

Development of the cost model is built on allocating or attributing factory costs and overheads to manufacturing technologies. Therefore, initially it is important to understand the current accounting systems in the relevant business, and to learn about what data are available. The prime element of data is likely to be a copy of the management accounts with a breakdown of all the cost elements for the most recent financial period (see step 2). With limited timescales it will be unjustifiable to study each cost element in detail, so a Pareto analysis will be useful for highlighting the major elements and providing a priority listing (see Fig. 4.5).

The major cost elements should then be classified into those which can be readily allocated to technologies, e.g. direct labour, depreciation, and those which have no obvious allocation basis, e.g. corporate charges, telephones.

For those readily allocated, bases for cost allocation, or cost drivers, need to be identified with advice from the finance manager. Some examples are given in Table 4.3.

Having selected a basis, data will need to be collected and analysed in order

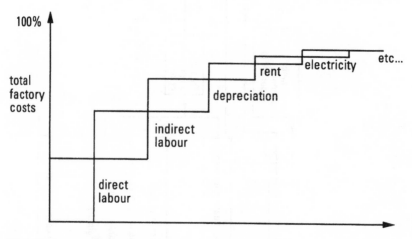

*Fig. 4.5 Pareto analysis of cost elements*

*Table 4.3 Example cost allocation bases*

| Cost element | Example allocation basis | Comments |
|---|---|---|
| direct labour | people | take account of pay differentials |
| indirect labour | how people spend their time | interview key people/ managers |
| Depreciation | | |
| - plant | items of plant allocated to technologies | check manufacturing routes and machine loadings |
| - buildings | buildings allocated on floorspace | |
| rent | floorspace | |
| rates | floorspace | |
| electricity | kilowatt hours | consult works engineering |

to derive how this basis allocates cost to each technology. Thus, element by element, a cost profile for each technology can be developed.

However, the Pareto analysis will suggest a law of diminishing returns, and some costs, although large, will have no obvious allocation basis. Potential to allocate may be no more than 50% of the total costs (excluding materials). It is useful to focus on the more variable cost elements, for example maintenance or tooling, since:

(i) the make or buy decision will affect these immediately;
(ii) comparison with supplier rates is usually dominated by the variable costs since some fixed costs remain even if the technology is eliminated.

Take care in deciding what is considered a fixed/variable cost element over what time frame. In theory, all cost elements are ultimately variable in the long term, i.e. what is fixed over a one year horizon may become variable over five years. Input from the finance function is useful here.

Judgement will be needed in each case to determine where to draw the line in terms of cost element allocation. A significant percentage (e.g. 40% to 50%) of the costs will probably have to be allocated on a pro rata basis (i.e. remaining costs apportioned in the same ratios as the allocated costs).

The total acquisition costs (TAC) including direct material spend and internal costs need to be separately identified as they relate to bought out parts rather than internal technologies. Thus for the purposes of building the cost model, these can be treated as an additional technology. This will enable the bought out costs, and indeed supplier quotes, to be supplemented by a percentage material surcharge to reflect these costs.

For each technology, a unit cost needs to be agreed, e.g. cost per standard hour produced or cost per batch produced, in order to determine the cost rate. This is done by matching the total technology cost to the number of units produced in the corresponding time frame.

A standard spreadsheet package such as *Lotus 1-2-3* or *Excel* should be adequate for the cost model, as illustrated by Fig. 4.6.

| Cost Element | Business Total (£) | Allocation Basis | Technology | | TAC |
|---|---|---|---|---|---|
| | | | T1 | T2....T17 | |
| Indirect labour | | | A1 | A2....A17 | AC1 |
| Direct labour | | | B1 | | AC2 |
| Depreciation | | | C1 | | AC3 |
| HQ charges | | | D1 | | AC4 |
| ↓ | | | | | |
| | Total cost per tech or TAC | | | | |
| | Cost per unit (std hr, batch) | | | | |

*Fig. 4.6 Example structure of cost model*

## Hints and tips

- use historical accounting data as these will have more detail to support them than budgets/forecasts;

- take care in considering which costs are fixed or variable over which timescale;

- continuous and strong input from the finance function is essential as this department will eventually own the model.

# Step 7: architecture/technology relationship

## *Objective*

To establish the link between manufacturing technologies, subsystems and parts families (TSPs)*.

To determine the manufacturing technology content of the subsystems and parts families.

## *Approach*

In reality the decisions on whether to carry out particular manufacturing technologies in house, and whether to make or buy particular parts, parts families and subsytems, are intimately connected. The decision to outsource a manufacturing process impacts all the parts which use that process, with significant consequences for process routings and control. On the other hand, transfer of a certain parts family out of the firm could leave some manufacturing processes nonviable for reasons of cost or capacity, with damaging consequences for other essential components dependent on those processes.

For this reason we make a thorough analysis of the technology content of the product subsystems and parts families. One way to imagine this relationship is as a cube, with axes of manufacturing technologies, subsystems and parts families, see Fig.4.7. The unit of measure of the technologies is normally standard hours. For particular processes such as heat treatment or plating, it may be appropriate to consider batches, as discussed in step 6.

Initial assessment can be in terms of one single product, decomposed by means of its architecture to reveal the dependence on individual manufacturing technologies. This then provides the basis for factoring up total manufacturing technology requirements at different product mixes and levels of output.

The consequences of sourcing options can then be explored with the help of matrices derived from the cube: subsystems versus parts families, subsystems versus manufacturing technologies and parts families versus manufacturing technologies; for an example, see Table 4.4. This simplified example is given for illustration purposes only; the complete version would contain many more technologies.

These matrices help to focus on both important areas and anomalies in terms of technology content and parts family volume. They help to frame

---

* For convenience, because they are frequently discussed together as a group, manufacturing technologies, subsystems and parts families will be referred to as TSPs

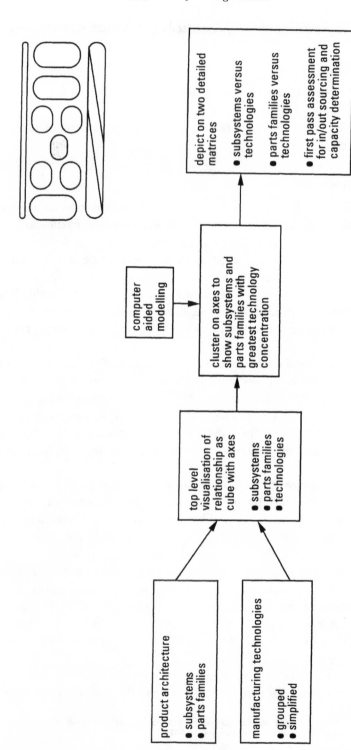

*Step 7 Architecture/technology relationship*

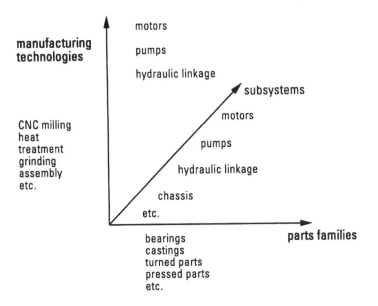

*Fig. 4.7 Manufacturing technology/subsystem/parts family relationship*

Table 4.4: *Example parts family/manufacturing technology matrix*

| Parts families | coil winding | pressing | CNC turning | conventional turning | milling |
|---|---|---|---|---|---|
| **Manufacturing Technologies (std hours)** | | | | | |
| Motors | 32 | 20 | 15 | 18 | 5 |
| Pumps | 0 | 12 | 11 | 7 | 6 |
| Hydr links | 0 | 6 | 12 | 8 | 7 |
| Total | 32 | 38 | 38 | 25 | 18 |

possible solution options for in and outsourcing, and they are useful in capacity determination. In deciding whether technologies are viable in or out, upper and lower limits on the capacity can be shown on the matrices.

The amount of data to be handled can be very large and in nearly all cases computer modelling will be necessary. This can be done with a personal computer spreadsheet.

## Hints and tips

- BOMs (bills of materials) and process routings provide most of the data, together with sales volumes, product architecture and manufacturing technology analysis;

- the cube model is mainly to build awareness of the links; detailed quantification of the technology content of parts families and subsytems is best shown on the matrices.

# Step 8: manufacturing technology assessment

## *Objective*

To assess the comparative importance of the technologies to the business and the competitiveness with which they are deployed, indicating any known trends.

To position the technologies on the assessment matrix and to form an initial view of the possible strategies for each technology.

## *Approach*

The assessment of technology competitiveness and importance is carried out largely by reference to the key success factors (KSFs) for the business. There are a number of generic KSFs for manufacturing industry (cost, quality, delivery and flexibility) which can be adapted for the specific conditions of the business under review.

Derivation of KSFs and the associated elements and drivers has been described in step 1. The question now is how the competitiveness and importance of the technologies can be assessed and their connection to the key success factors for the business.

### *8.1 Competitiveness*

Our competitiveness in the use of a technology can be assessed through a number of output measures. These will vary with the technology, but will typically be cost (per standard hour or unit), quality levels achieved (surface finish, tolerance, repeatability), time factors (set up time, throughput time) and flexibility factors (strength in depth, numbers of skilled people, ability to deploy the technology against varying demand). These are the familiar KSF groupings of cost, quality, delivery and flexibility. The overall assessment of competitiveness of the technology in a business will weight the contribution of the different component measures of competitiveness by the KSF weightings derived in step 1.

Competitiveness is always a comparative measure; the comparison is with what other users of the technology are achieving against these output measures, no matter in which field they are being applied. We should cast the net widely in looking outside for best practice in the use of technologies,

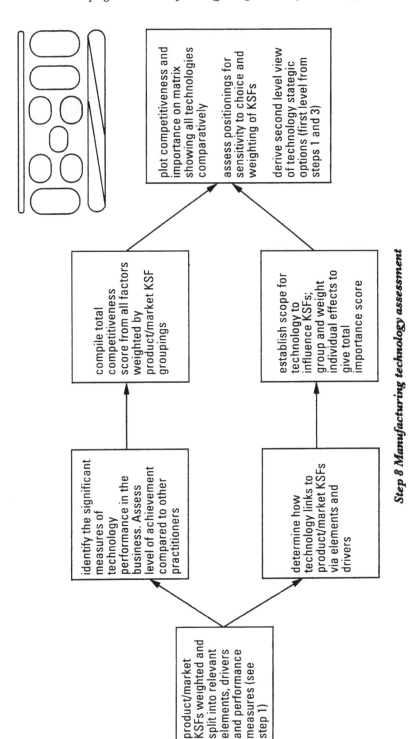

*Step 8 Manufacturing technology assessment*

as there may be new ideas to be picked up from unrelated fields. Substitute processes and technologies may even emerge during this external scan. However, given that the constituents of an overall competitiveness score have to be a realistic set, the comparison will normally be with others using the technology for similar purposes.

An example of how the competitiveness score for a particular technology, CNC milling, is derived is shown in Table 4.5. This is followed by an explanation of the various steps involved.

*Table 4.5 Derivation of competitiveness score for CNC milling*

| KSF | Measure | Best practice | Firm level | Weighting (0-1) | Score (0-5) | Total |
|---|---|---|---|---|---|---|
| Quality (wtg 0.2) | scrap | 10 ppm | 1% | 0.1 | 1 | 0.1 |
| | surface finish | no further treatment | finishing processes | 0.1 | 3 | 0.3 |
| Cost (wtg 0.4) | cost/ person hr | £35.00 | £56.00 | 0.4 | 3 | 1.2 |
| Delivery (wtg 0.3) | lead time | 1 day | 2 days | 0.1 | 3 | 0.3 |
| | schedule adherence | 95% | 80% | 0.2 | 2 | 0.4 |
| Flexibility (wtg 0.1) | multiskilled workforce | fully flexible | high flexibility | 0.1 | 4 | 0.4 |
| | | | | | | 2.7 |

Technology: CNC milling

The overall weighting of the KSFs in column 1 has been derived in step 1 for a particular product/market group. Column 2 shows the individual measures of technology performance which have been selected as relevant to the business, and where we can make comparisons with other practitioners. Columns 3 and 4 record best known performance against these measures compared with our own level of achievement. The weightings in column 5 are the split of the generic weightings in column 1 to each of the selected measures. Column 6 shows the score of our performance compared with external practice on a scale of 0 to 5, and the final column gives the total weighted score. This process of comparison and score compilation should be carried out as a group, so that a balanced point of view is reached.

The final score will be plotted on the competitiveness axis of the competitiveness/importance matrix on a scale of 0 to 5. In the given case of CNC milling the company has emerged as midranking in terms of competitiveness.

## 8.2 Importance

A technology's ability to influence the business KSFs is the indicator of its importance. For example, we and the competitors may be using the same manufacturing technology at similar levels of performance. In terms of competitiveness we shall score neutrally. However, if there is scope to make variations in our performance which will significantly impact one or more of the KSFs, then the technology is important to our business in comparison with another technology where there is no scope for change.

This importance may arise from the technology being a large contributor to a highly significant KSF, for example cost, even if there is only small scope for improvement. Or, conversely, the technology may only have limited impact on the KSF, but with huge scope for improvement. It is the overall size of the potential variation which determines importance. A critical technology that is operating on the limits of performance with consequent risk of failure and KSF degradation is similarly important.

The final result depends on the sum of the potential effects of the technology on all the weighted KSFs. This process is shown in Table 4.6. The measures now derived from the generic KSFs are the main ones that the customer will actually assess. There is also some work to be done by the project team before this table can be completed, and that is to determine the potential that each technology has to affect these measures. This results in the comparative score on the scale 0 (no impact) to 5 (very high impact), which is recorded in column 4.

*Table 4.6 Derivation of importance score for CNC milling*

| KSF | Measure | Weighting (0-1) | Score (0-5) | Total |
|---|---|---|---|---|
| Quality | % defects | 0.1 | 4 | 0.4 |
| (wtg 0.2) | features/ specification | 0.1 | 2 | 0.2 |
| Cost (wtg 0.4) | unit cost | 0.4 | 4 | 1.6 |
| Delivery | PIP lead time | 0.1 | 1 | 0.1 |
| (wtg 0.3) | schedule adherence | 0.2 | 2 | 0.4 |
| Flexibility | model changeover time | 0.05 | 2 | 0 1 |
| (wtg 0.1) | ability to accomodate load variation | 0.05 | 2 | 0.1 |
| | | | | 2.9 |

Technology: CNC milling

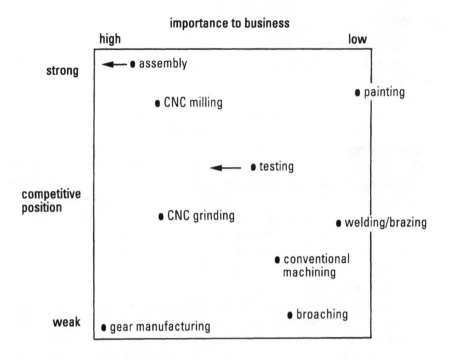

*Fig. 4.8 Example matrix positioning—technologies*

In conclusion the competitiveness/importance matrix is drawn showing all the technologies, Fig. 4.8.

If particular shifts in a technology's competitiveness or importance are anticipated because of known factors, these forthcoming shifts can also be shown on the matrix. This could arise from changing market conditions or technological breakthroughs. For example, in the case shown in Fig. 4.8, assembly and testing are becoming more important to the business.

In addition, by varying the weightings and composition of the KSFs, it is possible to carry out some sensitivity analysis. This is particularly useful for putting some limits on uncertainty. If sufficient information can be gathered, it is also useful to position suppliers on the matrix; this applies particularly when they are providing individual technologies.

The matrix begins to suggest some priorities for make-in/buyout when compared with the generic strategies matrix in Fig. 4.9.

## Hints and tips

- assessment of competitiveness is difficult and time consuming; there is a high volume of work and data collection should start early;

**importance to business**

| | high | medium | low |
|---|---|---|---|
| **strong** | continue to invest maintain capability | consolidate keep pace | licence/joint venture, reduce investment, or capability may open new market opportunities |
| **neutral** | invest develop | partnership | stop outsource |
| **weak** | initiate R&D examine for investment or cease, find comaker | partnership | stop monitor sell/licence design out find commodity supplier |

**competitive position** (row label, left side)

*Fig. 4.9 Generic strategies suggested by the technology matrix*

- information can be gathered from interviews (in house and with suppliers), perception questionnaires, sample quotes, trade associations, best practice indicators from industry and academic sources;

- check on supplier quotes: hours x rates = price?

- compare internal operation times with external to validate hourly rates;

- as each technology is analysed, keep a record of any constraints specific to that technology;

- the position of painting in Fig. 4.9 as being of low importance but with strong capability has emerged at several sites; where this technology is used in association with forming an outer case it may still be appropriate to consider outsourcing both activities together, rather than looking for ways to exploit it further;

- carry out sanity checks on the outputs – refer back to first pass greenfield scenario and manufacturing in the context of the whole business;

- the scaling of the matrix is arbitrary; the project team can set the scales with which they find it most convenient to work.

# Step 9: product architecture assessment

## *Objective*

To assess the comparative importance to the business and competitiveness of the subsystems and parts families making up the products. To identify trends and form an initial picture of the possible strategies for each item.

## *Approach*

The same approach for assessing importance and competitiveness via the impact on KSFs is used as in step 8. The project team will need to take a view as to how far to proceed with this analysis and possibly limit it to major parts families or subsystems where a question over sourcing emerges from a first pass analysis. The links between product architecture and customer requirements were made initially in step 4 (see also Chapter 7). This provides a useful entry into considering how the various subsystems and parts families can be assessed for competitiveness and importance.

### *9.1 Competitiveness*

Competitiveness of susbsystems and parts families is assessed through measures appropriate to the item under consideration. Cost and performance against specification are likely to be critical, together with delivery lead time and reliability. Where items are currently being outsourced some of these aspects will relate to the capability of a supplier in the context of the wider supply base, which can be considered as a measure of how good the business is at sourcing these items. The overall competitive position is established by weighting the various measures with product/market KSFs derived in step 1 and forming a total score. An example for an externally sourced parts family, ball screws, is shown in Table 4.7.

Some of the measures, for example of quality, would need to be defined in more detail and appropriately for the product under consideration. This Table is an example for comparison with the evaluation of technology competitiveness in step 8.1.

Through this analysis we have emerged as competitively weak in the field of the parts family ball screws, in the context of our product/market.

The analysis applies equally to items sourced inside or outside the firm. However, when considering outsourced items, the acquisition costs should be included before any internal comparison is made. If we find a competitive problem with an important outsourced item, we shall have to consider alternative sourcing strategies.

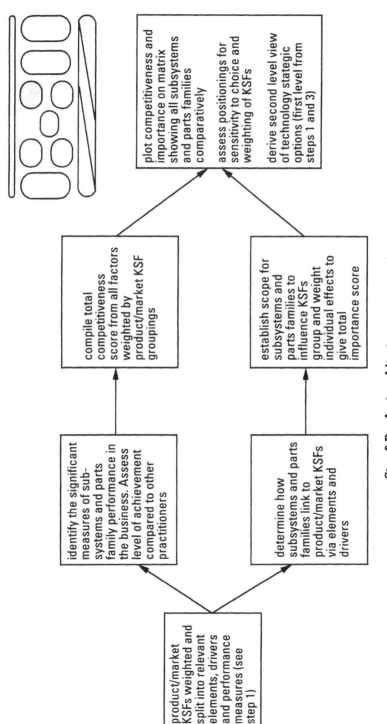

*Step 9 Product architecture assessment*

plot competitiveness and importance on matrix showing all subsystems and parts families comparatively

assess positionings for sensitivity to choice and weighting of KSFs

derive second level view of technology stategic options (first level from steps 1 and 3)

compile total competitiveness score from all factors weighted by product/market KSF groupings

establish scope for subsystems and parts families to influence KSFs group and weight individual effects to give total importance score

identify the significant measures of sub-systems and parts family performance in the business. Assess level of achievement compared to other practitioners

determine how subsystems and parts families link to product/market KSFs via elements and drivers

product/market KSFs weighted and split into relevant elements, drivers and performance measures (see step 1)

*Table 4.7 Derivation of competitiveness score for ball screws*

| KSF | Measure | Best practice | Firm level | Weighting (0-1) | Score (0-5) | Total |
|-----|---------|---------------|------------|-----------------|-------------|-------|
| Quality (wtg 0.2) | composite of measures: torque, backlash, etc. | top specification | top specification | 0.2 | 3 | 0.6 |
| Cost (wtg 0.4) | cost/item | £500 | £600 | 0.4 | 1 | 0.4 |
| Delivery (wtg 0.3) | lead time | JIT | six weeks | 0.15 | 1 | 0.15 |
| | delivery quantity | | minimum quantity/five | 0.15 | 1 | 0.15 |
| Flexibility (wtg 0.1) | range of types available | yes | difficult to change order | 0.1 | 1 | 0.1 |
| | | | | | | 1.4 |

Parts family: ball screws

*Table 4.8 Derivation of importance score for parts family ball screws*

| KSF | Measure | Weighting (0-1) | Score (0-5) | Total |
|-----|---------|-----------------|-------------|-------|
| Quality (wtg 0 2) | %defects | 0.1 | 5 | 0.5 |
| | product specification | 0.1 | 5 | 0 5 |
| Cost (wtg 0.4) | unit cost | 0.4 | 5 | 2 0 |
| Delivery (wtg 0.3) | PIP lead time | 0.1 | 4 | 0.4 |
| | schedule adherence | 0.2 | 5 | 1.0 |
| Flexibility (wtg 0.1) | model changeover time | 0.05 | 3 | 0 15 |
| | ability to accomodate load variation | 0.05 | 3 | 0.15 |
| | | | | 4.7 |

Parts family: ball screws

## 9.2 Importance

The subsystem or parts family's ability to influence the business KSFs is again the measure of importance. An example derivation is shown in Table 4.8, again for the parts family ball screws. Here, the particular measures which have been used within each KSF are the same as Table 4.7 for ease of comparison. The project team would carry out some additional work to understand the link of the parts family to these (or other appropriate) measures and their scope to influence them.

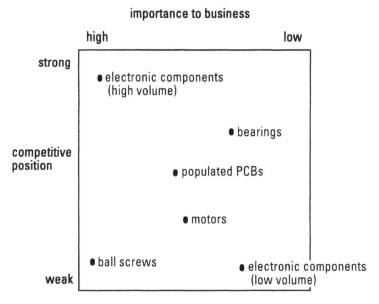

*Fig. 4.10 Example matrix positioning—parts families*

In conclusion, the competitiveness/importance matrix is drawn first for subsystems and secondly for parts families. An example is shown in Fig. 4.10.

## *Hints and tips*

- try to eliminate as much subjectivity as possible and watch for unintentional bias, particularly in assessing importance – there is a tendency to overestimate our own competitive position;

- avoid paralysis by analysis, make a rough cut assessment first, where possible, to determine where more detailed analysis would be useful; it is not necessary to analyse all technologies, subsystems and parts families to the same degree.

## Step 10: decision support modelling

### *Objective*

To develop (computer) models to show the effect of strategy options on the business measures.

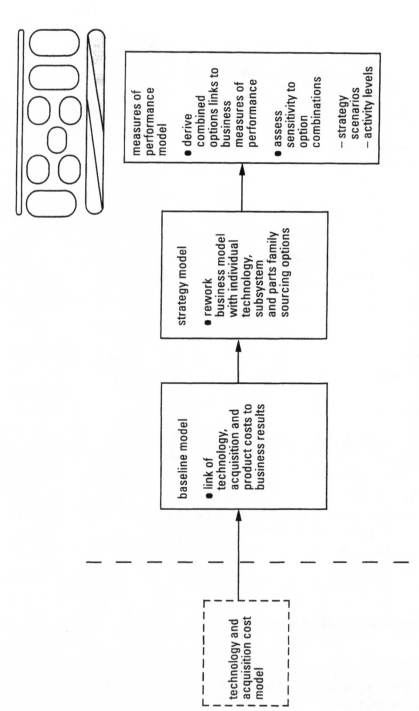

**baseline model**

- link of technology, acquisition and product costs to business results

**strategy model**

- rework business model with individual technology, subsystem and parts family sourcing options

**measures of performance model**

- derive combined options links to business measures of performance

- assess sensitivity to option combinations

  – strategy scenarios
  – activity levels

technology and acquisition cost model

*Step 10 Decision support modelling*

## *Approach*

Modelling the effects of different sales levels and strategy choices builds on the work done in creating the technology cost model (step 6). In addition to this model we now build three related models: the baseline model, the strategy model and the measures of performance model. The individual purpose and structure of each of these is described below. The detail of the particular cost elements, variables and measures of performance will vary from business to business. However, in order to clarify the use of the models, examples are given from a specific case, and some typical figures and assumptions are quoted.

### *Baseline model*

This provides projections for the main cost elements identified by the cost model for each internal technology based on the forecast sales scenarios. It assumes no constraints on space, personnel or investment and no movement of technologies, subsystems and parts families from the current situation. The results of these projections should be broadly in line with existing business forecasts – this should be checked to make sure that the model is soundly based.

An example structure of the baseline model is shown in Fig. 4.11. Some of the variables may require additional definition:

*Projected standard hours with experience effect*—hours are assumed to be proportional to forecast sales based on knowledge of current standard hours relating to current sales. Differences in the mix of product group volumes and their differences of internal technology content are accounted for in the calculation by the use of ratios.

*Projected standard hours with experience effect*— hours are assumed to reduce with the cumulative experience gained in production, based on methods improvement. Typically standard hours reduce to 70 – 80% of the original standard time for a doubling of production volume. Production volume is assumed to be proportional to sales volume.

*Labour productivity*—based on a comparison of clocked and standard times. Projected labour efficiency is assumed to improve by 15% with each doubling of standard hours, due to increasing operator speed.

*Projected direct hours*—calculated from projected standard hours and labour productivity.

*Projected change in indirect staff* related to the change in direct staff, but with the proportion indirect/direct decreasing as production volume increases.

### Strategy model

This incorporates the strategy options to modify the standard hours to be worked. It then calculates the staff, space, electricity, work in progress and changes in bought out costs as a result of these strategy choices. The structure of the model is shown in Fig. 4.12. Additional explanation is as follows:

*Projected standard hours with experience effect after strategy*—hours are eliminated or phased out over time as a technology, subsystem or parts family is outsourced, or increased if they are brought in house. The subsequent reduction in associated technologies must also be accounted for.

*Projected work in progress (WIP) costs*—assumed to be proportional to projected standard hours.

*Projected supplier rates* — assumed to decrease with increase in order volumes. A reduction based on a 20% reduction for five times the current order volume is assumed. Order volumes are assumed to be proportional to sales volume. Check assumptions with suppliers.

*Projected change to bought out costs*—as a result of strategy choices, changes to the standard hours multiplied by the bought out supplier rate gives the change to bought out costs.

### Measures of performance (MOP) model

The MOP model gives the key business performance measures over the next planning period (typically three to ten years) for the sales scenarios. It then shows the changes to these measures which result from the strategic plans for each of the sales scenarios. The structure of the model is shown in Figs. 4.13 and 4.14. It makes use of data from the strategy model, and also information concerning investment, divestment, depreciation, creditors and stock.

## Hints and tips

- involvement of the finance department is critical;
- get ownership by the finance manager if possible;
- understand the basis of existing forecasts; sales, costs, mix, etc;
- use the same base from which to make projections;

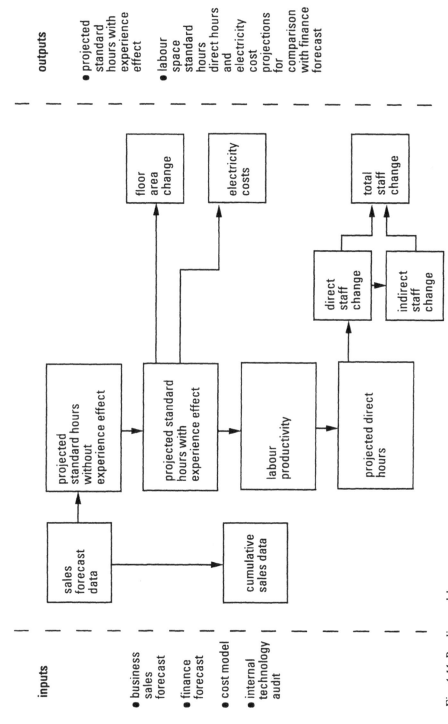

**inputs**
- business sales forecast
- finance forecast
- cost model
- internal technology audit

**outputs**
- projected standard hours with experience effect
- labour space standard hours direct hours and electricity cost projections for comparison with finance forecast

sales forecast data

cumulative sales data

projected standard hours without experience effect

projected standard hours with experience effect

labour productivity

projected direct hours

floor area change

electricity costs

direct staff change

indirect staff change

total staff change

*Fig. 4.11 Baseline model*

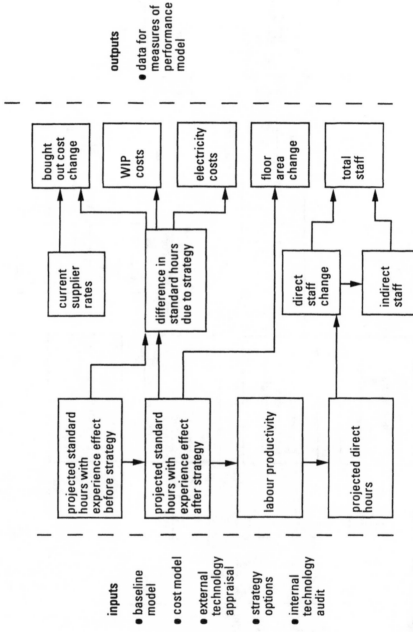

**outputs**
• data for measures of performance model

**inputs**
• baseline model
• cost model
• external technology appraisal
• strategy options
• internal technology audit

*Fig. 4.12 Strategy model*

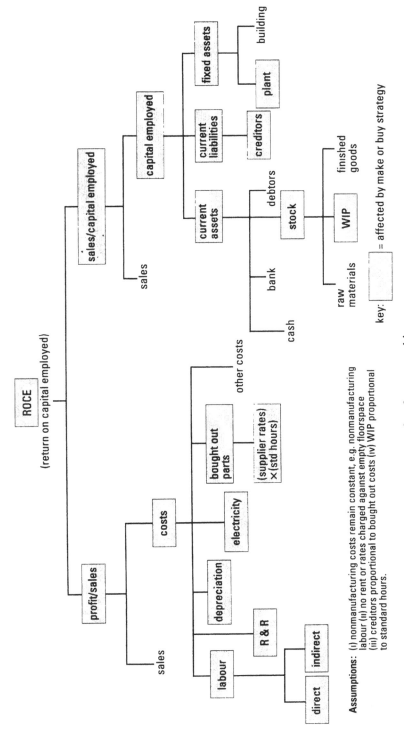

**Assumptions:** (i) nonmanufacturing costs remain constant, e.g. nonmanufacturing labour (ii) no rent or rates charged against empty floorspace (iii) creditors proportional to bought out costs (iv) WIP proportional to standard hours.

*Fig. 4.13 Identification of the variables in the measures of performance models*

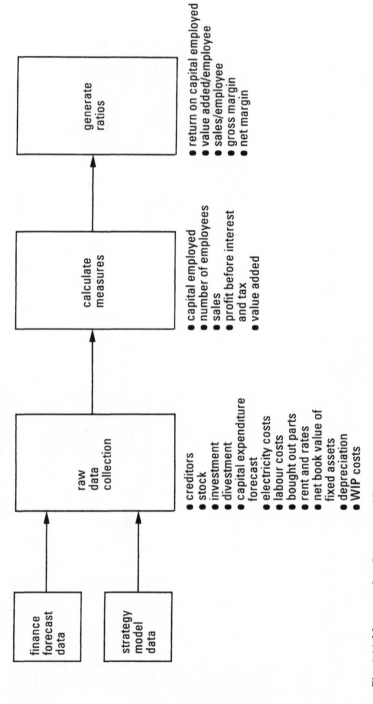

*Fig. 4.14 Measures of performance model*

- spreadsheets are adequate for model building;

- minimise computing complication.

## Step 11: evaluate technology/subsystem/parts family options

*Objective*

To identify and evaluate the range of sourcing options (make-in/buyout) for the technologies, subsystems and parts families (TSPs).

*Approach*

By this stage in the methodology the realistic make/buy options are beginning to become clear. They will have developed from the initial first pass ideas of the greenfield scenario, through the more detailed analysis of the competitiveness/importance matrices. It may even be that for some technologies, subsystems and parts families the sourcing decision is now effectively made. For these TSPs there will be clear and overriding reasons why they should be retained inside the business, or sourced from outside. However, for the majority there will still be a range of possible options which need to be evaluated systematically. These include:

- retain in house, invest as required;

- exploit more fully through subcontracting or new product design;

- bring in house immediately;

- bring in house progressively over time;

- monitor with a view to bringing in house later;

- monitor with a view to buying out later;

- leave out;

- buy out progressively over time;

- buy out immediately;

- design out of products.

For each TSP the preceding analysis will have narrowed the choice to a few possibilities within this range of options. In each of the options where a

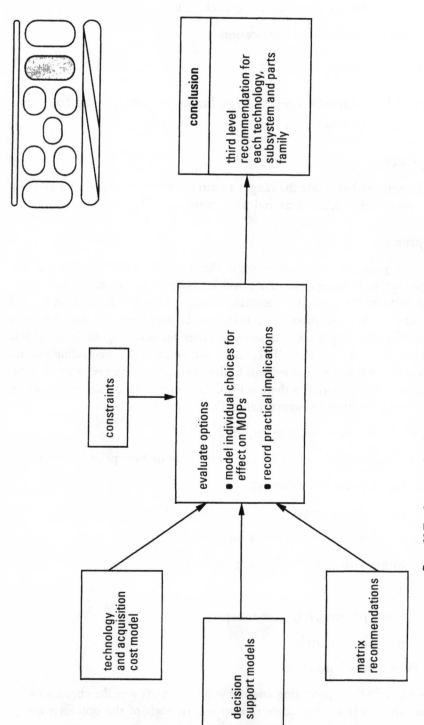

*Step 11 Evaluate technology/subsystem/parts family options*

supplier is involved, there is the question of the type of supplier relationship to be developed. The make or buy strategy is only intended to provide first indications for this; a more detailed evaluation is the subject of follow on strategic sourcing work. Analysis of the technologies, subsystems and parts families so far has largely been geared to the key success factors for the business. In an unconstrained world we would now develop a strategy which maximised our leverage on these factors, without reference to cost or the inherited circumstances. However, in practice there are some compromises to be made.

The next step in the process is first to identify all the constraints which exist and will be relevant, and then to consider each technology, subsystem and parts family in turn. The depth of analysis necessary is guided by the initial rough cut evaluation, the considerations of the greenfield option and the output of the matrix evaluation.

## Constraints

The individual constraints will be particular to the business, but can be categorised as follows:

*Company financial/performance measures*

- requirements for financial return (ROI, etc.);
- headcount limits;
- cash limits;
- stock targets.

*Strategic issues*

- the strategic issues derived from the SWOT analysis (step 1);
- company commitments to particular markets, products, locations;
- environmental and political factors.

*Operational aspects*

- mutual impact of transfers of technologies, subsystems and parts families in/out of the business (viability of what is left);
- implications for new product development;
- implications for skills and training.

Within these categories there could be many other criteria in addition to the examples given. A comprehensive listing is the first task in this step.

## Evaluation

Having established the criteria, the next task is to scan through all the technologies, subsystems and parts families against all the criteria, identifying the go/no go conditions. This means that for each item there could be one or more of the constraints determining whether the item should move inside or outside the firm, or cannot be moved from its current location. This process can be thought of as filtering the items past the constraints, although a final conclusion is not possible until all the effects are taken into account. The process should be carried out in the sequence technologies, then subsystems then parts families, in order to facilitate the evaluation of the effects of one on the other. As a result of this filtering process there will be a number of TSPs to be definitely kept inside the business and others definitely going out. However, the majority are still likely to be in the grey area between and a means of prioritising is now required.

A graphical approach to showing the results of this further analysis is useful as a way of visualising the individual items and their interaction. An example is shown in Fig. 4.15. In effect this is a compilation of all the factors covered in the methodology to date and indicates an overall make-in/buyout ranking for the TSPs.

The essential aspect to emerge from this step is the collective overview of the individual TSP options, with the associated ranking in terms of a make-in/buyout priority. The dependencies between the TSPs will be made explicit, leading to the identification of some viable combined options which could be implemented.

The decision support models may also be used to help with the assessment of the consequences on the business measures of performance of these individual TSP options. However, we are not concerned at this stage with fine detail modelling and financial projections, but rather the top level impact of individual TSP options. The full detail of the decision support models is better applied when realistic option combinations are being evaluated in the next and final step, 12.

## Hints and tips

- the possibilities of complicating this analysis abound - keep it simple!

- document decisions and conclusions so that they can be revisited;

- in proposing to move a technology in or out, watch process routes;

| technology | importance | competitiveness | impact on business MOPs | investment required | impact on other TSPs | impact on strategic issues | impact on new product development | comment | conclusion |
|---|---|---|---|---|---|---|---|---|---|
| technology a | 4.5 | 4.0 | ● | ◐ | ◐ | ◐ | ◐ | essential technology to retain and develop | keep in |
| technology b | 4.5 | 2.0 | ◐ | ● | ◑ | ◐ | ◐ | Although important, competitiveness has slipped Invest | keep in, finance permitting |
| technology c | 2.5 | 4.0 | ◐ | ◕ | ● | ● | ◐ | linkage to other tsp's and strong competitiveness mean we should maintain in house | keep in, keep pace |
| technology d | 2.5 | 2.5 | ◑ | ○ | ◑ | ◑ | ◑ | mediocre competitiveness and high investment requirement. Candidate for outsourcing | seek advantaged supplier with whom to cooperate |
| technology e | 1.0 | 1.5 | ◑ | ● | ◕ | ◕ | ○ | looming investment and low strategic value | look for best price supplier |
| technology f | 3.0 | 2.0 | ◑ | ◑ | ◑ | ◑ | ◑ | competitiveness requires improvement; develop or find partner | invest and develop |
| etc. . . | | | | | | | | | |

**key**:  importance:   0 – 5, low to high   competitiveness:  0 – 5, weak to strong   other factors:  ● strong reason to keep in   ○ no reason to keep in

*Figure 4.15 Assessment of individual technology options*

- get an independent reaction to conclusions;
- scoring is by nature judgmental - get a mix of opinion rather than aiming for precision;
- make frequent sensitivity checks.

## Step 12: develop strategy recommendations and implications

### *Objective*

To establish the optimal strategy, model the implications and define implementation requirements.

### *Approach*

The main activity of this step is to make explicit the practical consequences of the chosen strategy. This will be the combination of individual TSP options, which gives the optimal projected business measures of performance (with the help of the decision support models), together with a workable business design in terms of the manufacturing system and sourcing arrangements.

#### *Selection of the optimal strategy*

From the consideration of the individual TSP options in step 11, a number of possible combinations can now be hypothesised. Depending on the stringency of the criteria being matched, there may be only one viable combination. However, it is more likely that there will be two or three, which now need to be compared. This comparison is carried out with the help of the decision support models.

#### *Business measures of performance*

The first step in the process (step 1, *derive the business issues*), will have identified the particular measures of performance relevant to the business. These are then modelled by the decision support models described in step 10. Typical measures will be return on capital employed (ROCE), gross margin, cash flow, sales per employee and value added per employee. In selecting the chosen strategy and company structure these measures are projected over the planning horizon to identify the optimal result. This is then compared with

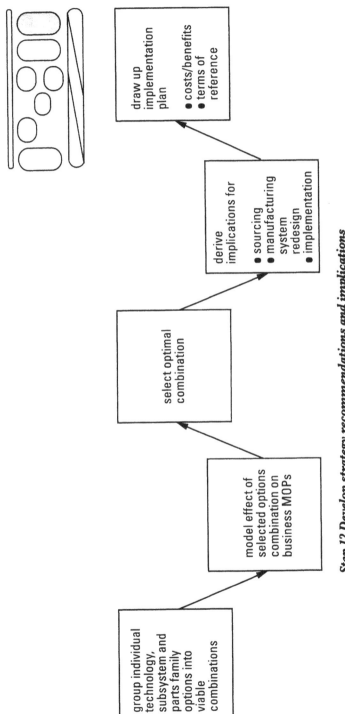

*Step 12 Develop strategy recommendations and implications*

the effect of continuing with the existing strategy and company structure, in order to provide a basis from which to measure future progress. An example is shown in Table 4.9. In addition, these projected measures provide the means to calculate the costs and benefits of the change of strategy.

An important task now is to carry out some sensitivity and risk analysis. This is to ensure that, as far as possible, the chosen strategy does not expose the company to potential damage. The effects of changes in levels of activity, product mix and costs can all be modelled; the consequences on the MOPs for each should be recorded, together with an assessment of the probability that such an event could occur. Escalation of particular supplier costs is a typical area of concern.

A top level view of the exposure to risk is given by calculating the break even point for the business. This is the activity level at which the business can just cover its costs. The sensitivity of the break even point to volume variation and actions to lower the break even point may be significant factors in strategy formulation. Consideration of the break even point may be particularly important in asset intensive businesses. A strategy to move to a more focused business can lower the break even point, making the company less vulnerable to the effects of volume variation . The calculation of the break even point is illustrated in Fig. 4.16

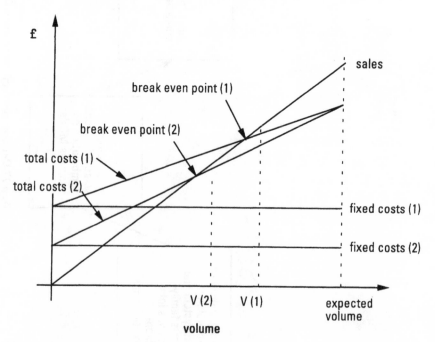

*Fig. 4.16 Calculation of the break even point*

Table 4.9 Business measures of performance – 10 year projection with and without proposed strategy

| | YEAR | 96/97 | 97/98 | 98/99 | 99/00 | 00/01 | 01/02 | 02/03 | 03/04 | 04/05 | 05/06 | 06/07 |
|---|---|---|---|---|---|---|---|---|---|---|---|---|
| Sales (£M) | | 33.6 | 41.6 | 34.4 | 40.8 | 56.8 | 66.4 | 87.2 | 114.6 | 117.6 | 133.1 | 148.8 |
| Employees | current plan | 652 | 656 | 698 | 728 | 790 | 796 | 862 | 936 | 984 | 1022 | 1058 |
| | with strategy | 652 | 614 | 654 | 682 | 678 | 670 | 712 | 750 | 778 | 796 | 774 |
| Capital | current plan | 15.4 | 16.1 | 17.1 | 19.4 | 20.4 | 23.2 | 27.1 | 27.6 | 29.4 | 31.6 | 31.6 |
| Employed (£K) | with strategy | 15.4 | 14.4 | 15.2 | 16.6 | 12.8 | 13.6 | 15.4 | 11.8 | 12.2 | 11.8 | 5.6 |
| Sales/ | current plan | 51.5 | 63.4 | 49.3 | 56.0 | 71.9 | 83.4 | 101.2 | 122.4 | 119.5 | 130.2 | 140.6 |
| employee (£K) | with strategy | 51.5 | 67.8 | 52.6 | 59.8 | 83.8 | 99.1 | 122.5 | 152.8 | 151.2 | 167.2 | 192.2 |
| Gross | current plan | 33.5 | 29.9 | 29.0 | 28.0 | 24.0 | 23.1 | 19.5 | 16.8 | 15.9 | 14.5 | 14.1 |
| margin (%) | with strategy | 33.5 | 30.0 | 28.1 | 27.6 | 22.7 | 22.8 | 19.1 | 16.1 | 15.3 | 14.0 | 12.8 |
| Net | current plan | 8.4 | 9.7 | 7.9 | 9.7 | 8.2 | 9.3 | 11.0 | 11.2 | 11.7 | 11.3 | 11.9 |
| margin (%) | with strategy | 8.4 | 9.8 | 6.9 | 9.3 | 6.8 | 9.1 | 10.6 | 10.5 | 11.2 | 10.9 | 10.9 |
| Value added/ | current plan | 23.5 | 26.7 | 25.0 | 27.5 | 28.5 | 31.0 | 34.8 | 38.0 | 38.7 | 40.1 | 42.6 |
| employee (£K) | with strategy | 23.5 | 27.5 | 25.7 | 28.2 | 30.5 | 33.5 | 38.5 | 43.1 | 44.3 | 46.3 | 51.8 |
| ROCE (%) | current plan | 18.3 | 25.2 | 15.9 | 20.5 | 22.8 | 26.7 | 35.6 | 46.8 | 46.8 | 47.6 | 56.3 |
| | with strategy | 18.3 | 28.4 | 15.7 | 22.9 | 28.5 | 36.6 | 50.2 | 63.5 | 69.7 | 72.3 | 85.2 |

This Figure shows the difference in break even points (V1 and V2) between two fixed/variable cost structures for the business under review, which both have the same total costs at the expected volume of operation.

A comparison of the projected performance between the option combinations, together with a review of the implementation requirements of each, will lead to the selection of the optimal strategy.

*Manufacturing*

The changes required in manufacturing capability should now be examined in more detail. There will be consequences for type and capacity of process, in some cases with complete manufacturing technologies, parts families and subsystems moving in or out of the organisation. In addition to the hardware considerations, the implications for numbers of people, skill profiles and manufacturing organisation need to be worked out. In cases where a move of technology is indicated, the age and condition of plant may largely determine when it is sold, renewed or replaced. The identification of the times at which these major shifts in resources occur is critical to ensuring adequate preparation. It is useful to refer back to the decision areas of manufacturing strategy (Fig.2.2), to make sure that the implications for each of these has been considered.

*Sourcing*

The indications of the preferred approach for the outsourcing of technologies, parts families and subsystems also need to be made more explicit. The conclusions from the analysis so far lead to implications for supplier relationships and supplier requirements. A detailed evaluation of these aspects is beyond the scope of the make or buy review, but depending on the terms of reference for the project and the scale of the business it could be useful to carry the analysis a little further.

A number of aspects are involved for each parts family:

- product challenge – how difficult is it to design and manufacture?

- sourcing challenge – how able are suppliers to meet requirements?

- annual spend.

These are scored 1-3 to give low/high influence and easy/hard difficulty. The parts families are then plotted on matrices of annual spend/product challenge and sourcing challenge/product challenge, Fig.4.17, which provide indications of the sourcing requirements and focus. This analysis is one level more detailed than has come out of the make or buy review so far,

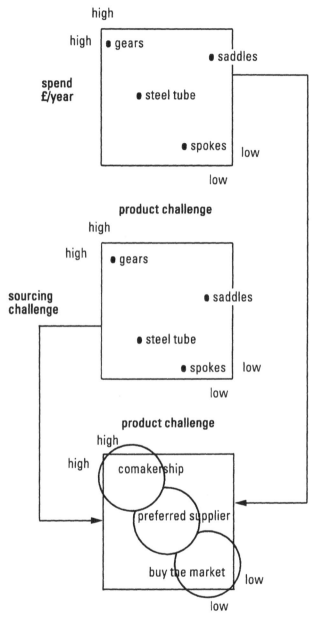

*Fig. 4.17 Guide to sourcing strategies*

but considerably more work would be required before changes to sourcing arrangements were made.

*Implementation plan*

The implementation plan is to a large extent a summary of the actions arising from the implications for sourcing and manufacturing. It provides the guide to executing the new strategy and indicates the timing and phasing of changes in technologies and parts families. Particular tasks and projects are defined, together with the necessary resources and expertise. The main shifts of resource should be charted over the full planning horizon, but with a detailed project plan for the first year. In any case, this should be in sufficient detail for work to begin on the implementation following management review and sign off of the proposed new make or buy strategy.

In total the implementation plan should cover the following aspects:

- communications – how, when and what to tell the people not so far involved;

- resource – if necessary setting up a team with roles and responsibilities;

- manufacturing

  —acquisition/disposal of plant
  —restructure groupings
  —adaption of planning and control systems
  —personnel training/skills/recruitment;

- sourcing

  —define strategies
  —supplier selection and screening (for priority technology and parts family groups)
  —timing of changes
  —structuring the supplier relationships;

- organisation

  —restructuring manufacturing support functions
  —restructuring the purchasing function
  —revisit/follow up business strategic issues.

## Hints and tips

- double check inputs and outputs of models;

- check that the effects of final recommended manufacturing and sourcing changes have been modelled;

- executive conviction and ownership is vital for further implementation;

- management of communication outside the project team is critical to the implementation.

## Chapter 5
# Implementing and maintaining the strategy

## 5.1 Importance of a phased approach

The phased approach to the strategic review, which is the basis of the formulation of a new make or buy strategy, has been described in Chapter 3. The importance of the phases is that they provide the means to divide the work into managed stages, with review points at the end of each stage. These reviews will usually take the form of a management workshop, at which findings and proposals are presented. This provides the opportunity to achieve management validation of the data and reasoning, and most importantly, ownership of the conclusions and actions, with commitments to move on to the next stage.

The review is conducted as a project built around these phases, the setup

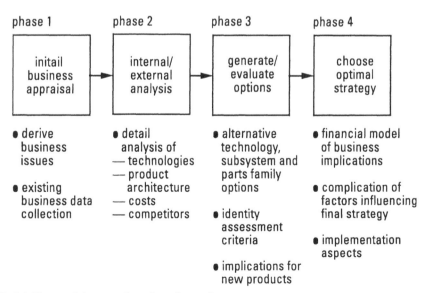

*Fig.5.1 Phases of the strategic make or buy review*

and management of which is critical to the success of the review. In addition, some organisations find it useful to involve an outside agency to support the execution of the project. This can take the form of internal or external consultants, or some form of facilitation. Detailed prior knowledge of the approach described is not essential, depending on the precise role fulfilled by the agency.

## 5.2 Project setup and management

Clear objectives, a sound working structure and regular reviews are essential to the success of any project. A strategic make or buy review is so wide ranging in its scope, with implications for so many areas of the business, that it is particularly important to make most effective use of the project team resources. This section outlines the requirements of project preparation and control, together with an example project specification.

### General management prebriefing

Thorough preproject discussion* with the general manager of the business is essential. This person is the ultimate project owner, and there are several areas which need to be dealt with before work begins. These include:

- what business problem is the general mnager trying to solve?
- the potential implications of the work for the business;
- the scale of the effort required by all concerned ;
- involvement of the whole executive team ;
- how to maintain the security of the project team proceedings;
- how and when to communicate to the wider workforce ;
- if day to day ownership is to be delegated, to whom;
- skills and experience required from team members.

Of these the most critical factors are the need for an overall strategic direction set by top management, and the requirement to achieve management team concensus. The first is essential, because without it the project team will find it impossible to assess the comparative merits of the options which they generate. The second is also very important, if the whole management team is to endorse the strategy which emerges from the review and see it through to implementation.

When these aspects have been resolved it is possible to go ahead and draw up the project terms of reference.

---

* Who should be involved in this discussion depends on how the project is managed and whether a steering committee is established. At the minimum it should be between the site general manager and the project manager

An example of the project terms of reference is given below. It comprises both a statement of the overall aim of the project and the more detailed individual objectives, or deliverables, for the team to work to. A clear and explicit statement in this form has two major benefits. It gives the project team distinct objectives, and it ensures that the members of the management team commissioning the work are fully aware of the scale and complexity of the task. As a result they should be better prepared to commit the necessary resources.

## Example aim

To define a make or buy strategy for the business in question. This is to provide both long and short term (tactical) guidelines for business decision making as part of the overall manufacturing strategy. It will include consideration of the impact of the strategy on nonmanufacturing areas, in particular product design.

## Example objectives

1. Determine the appropriate level of vertical integration, examining opportunities both upstream and downstream on the value added chain.
2. Assess the position of maufacturing compared with other activities in the business (for example design and development, sales and support, etc.) in terms of its competitiveness and importance.
3. Establish a cost model to determine the real costs of manufacture, addressing the allocation of overheads to manufacturing process technologies, and enabling comparisons to be made with external technology sources.
4. Recommend improvements to the way sourcing decisions are taken in the product introduction process, and identify the appropriate extent and timing of supplier involvement.
5. Produce recommendations on the approach to be taken in sourcing the technologies, subsystems and parts families which are to be bought out, and indicate the implications that the chosen strategy has on the sourcing organisation.
6. Identify implications for the development of the business and manufacturing organisation.
7. Determine the impact that the introduction of the above make or buy strategy will have on business performance, as measured by the site general manager. Measures should include the break even point for the business.
8. Recommend the next steps needed to implement the strategy, including the approximate scale, duration and resource requirements for the work.
9. Produce a user manual for the management team to help guide decision making during implementation and to help any further sensitivity analysis.

## Project management and control

The main work of the project is carried out by the task force which acts as an agent of the management team of the business, with full access to the necessary information. In order to provide a monitoring and guiding input to the project, detached from the day to day activities of the site, a steering group may be constituted to oversee the whole process from setting up of the terms of reference to ensuring that progress is maintained and deliverables achieved. The relationship of these three groups is shown in Fig. 5.2.

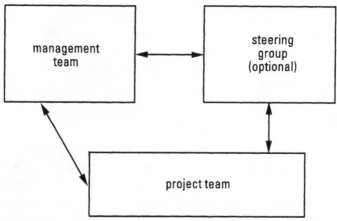

*Fig. 5.2 Project management structure*

The need for a steering group depends very much on the overall nature of the organisation, in particular the relationships between the local site and any corporate bodies, and whether external consultants or facilitators will be involved in the project. The steering group may then consist of local site management (the project owner), corporate and consultant organisation management if appropriate and any necessary technical expertise. The group should meet at least monthly following the workshop reviews which summarise and sign off each of the four main stages of the approach. The workshop reviews involve taskforce, management team and steering group and are decision points in terms of reviewing the analysis and recommendations to date, and direction setting for the next steps in the project. They are critical to joint ownership of the emerging strategy.

## Taskforce composition

The taskforce requires a full time project team of five or six people, representing the following functions or roles:

project manager

manufacturing systems engineer
manufacturing/manufacturing engineering
supplies/purchasing
design engineering
finance/accounts
analyst (graduate trainee or equivalent)

A blend of skills and experience is required within the team. Essential are business and product knowledge, engineering and manufacturing experience, knowledge of the supplier base, accounting knowledge and computing skills. Some thought must be given to the selection process for achieving this mix, together with the all important ability to look at the issues from a strategic perspective and not get trapped in the detail. This group works together full time, dividing the tasks between them. Daily quick planning and review meetings are held among the team, and more formal weekly reviews with the project owner from the management team. In addition, there may be regular (two to three weekly) reviews with a consultant/facilitator organisation, if one is used, and the full end of phase workshops described above.

## Timescale

The approximate timescale for a typical project, including three days for project team training, is ten weeks. The phasing of the activities is shown in Fig. 5.3. In scaling these activities for varying business conditions the important variables are single or multisite, product and process complexity, ease of access to the managment team, need for customer and supplier surveys, competitor benchmarking and the availability of data for the cost model. The typical example given above is drawn from the case of a single site business of medium size with two main product groups, seventeen manufacturing technologies, good access to the management team and full information available.

## Project team training

An example of the introductory training programme is shown below. The training has a number of purposes:

- common understanding of project requirements;
- team building;
- equipping the team with relevant skills and knowledge;
- establishing confidence in the methodology;

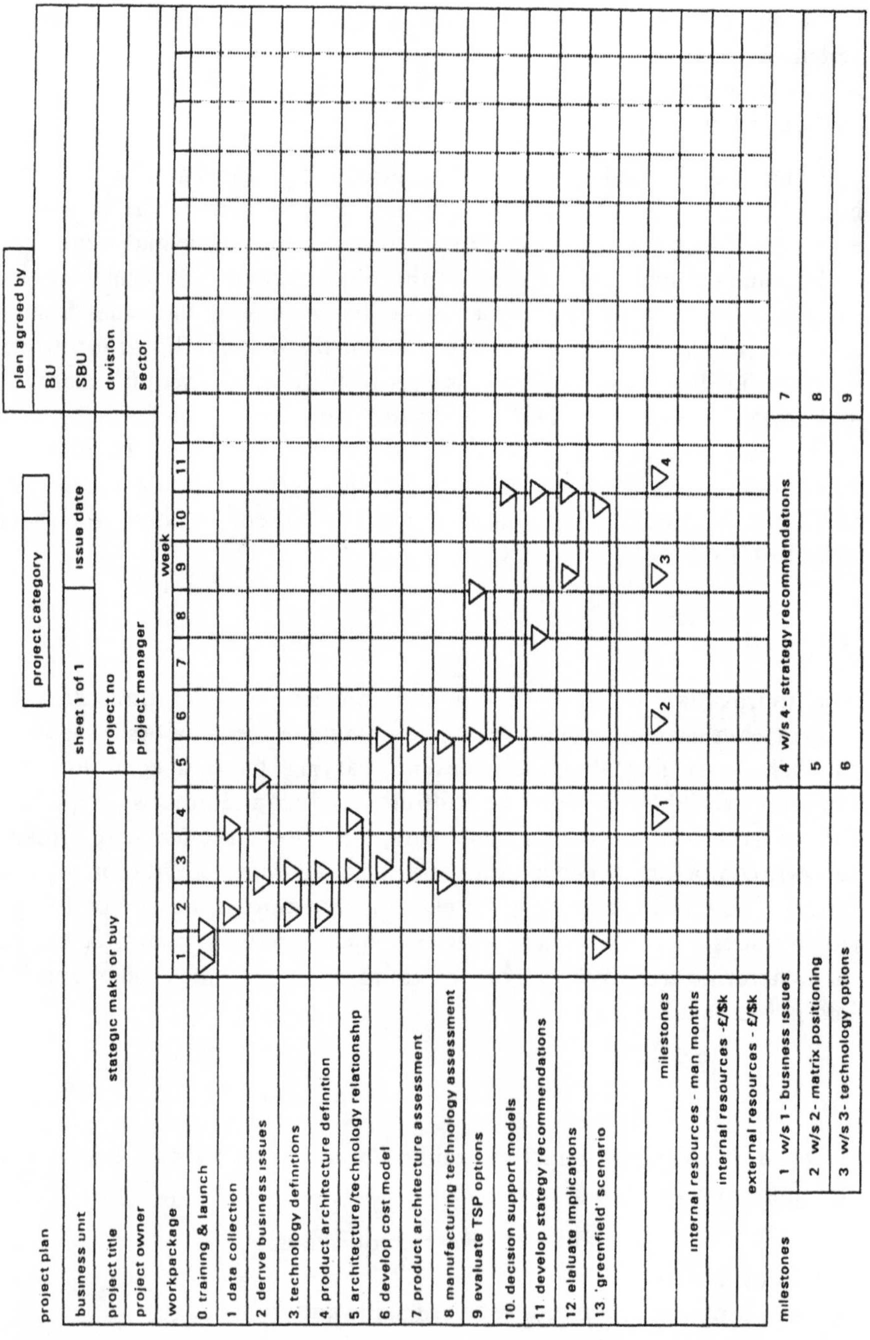

project plan

plan agreed by

| | |
|---|---|
| business unit | BU |
| project title    stategic make or buy | SBU   division |
| project owner | sector |

project category

sheet 1 of 1     issue date

project no

project manager

**workpackage** — week (1–11)

0 training & launch
1 data collection
2 derive business issues
3 technology definitions
4 product architecture definition
5 architecture/technology relationship
6 develop cost model
7 product architecture assessment
8 manufacturing technology assessment
9 evaluate TSP options
10. decision support models
11 develop stategy recommendations
12 elaluate implications
13 'greenfield' scenario

milestones

internal resources - man months
internal resources -£/$k
external resources - £/$k

milestones

1 w/s 1 - business issues
2 w/s 2 - matrix positioning
3 w/s 3 - technology options
4 w/s 4 - strategy recommendations
5
6
7
8
9

*Fig. 5.3 Activity phasing*

- developing and agreeing a detailed project plan;
- signalling the launch of the project.

This generic programme will need to be adapted to local requirements.

## 5.3 Generic project team training plan

| Session title | Provider/ trainer | Session objectives |
|---|---|---|
| **(Day 1)** | | |
| 1. Introduction and objectives | project manager | introduce the training programme and personal introductions |
| 2. Business strategy<br>&bull; strategy<br>&bull; markets, products, customers<br>&bull; key success factors<br>&bull; competitors/benchmarking<br>&bull; SWOTs<br>&bull; objectives, targets<br>&bull; other issues | general manager | present and discuss business environment and goals<br><br>draw out relevant issues |
| 3. Terms of reference | steering committee | introduce and discuss project specification |
| 4. Manufacturing strategy and vertical integration | invited expert | gain understanding of theoretical background to the subject |
| 5. Team building | project manager | exercises to promote working as a team and identify strengths/ weaknesses |
| **(Day 2)** | | |
| 6. Make or buy methodology<br>&bull; overview<br>&bull; positioning manufacturing<br>&bull; process technologies<br>&bull; product architecture<br>&bull; benchmarking/matrix positioning<br>&bull; cost modelling<br>&bull; strategy formulation and evaluation | project manager/ facilitator/ external expert | provide a route map for the approach<br><br>discuss important techniques |

| | | |
|---|---|---|
| 7. Case example | as above | gain understanding of real life example |
| 8. Overview of current costing systems and data | finance manager | understand data available and current systems in use |

**(Day 3)**

| | | |
|---|---|---|
| 9. Overview of strategic sourcing | invited expert | appreciate approach and objectives |
| 10. Make or buy issues for product introduction | invited expert | understand relationship between make or buy decision and product introduction process (PIP) |
| 11. Project management | project manager | agree approaches to project planning and control |
| 12. Project planning | team | plan project in detail, identifying timescales, milestones and workpackages |
| 13. Prepare for launch | team | prepare launch presentation* |

**Launch**

## 5.4 Maintenance of the make or buy strategy

Having devised a strategy which is optimal for the business goals, it is important to keep it up to date and abreast of a changing environment. This is best done by identifying the critical conditions which have shaped the strategy, and then monitoring them for any changes. The following is a list of the factors most likely to be important in terms of tracking for significant change from the conditions under which the strategy was developed:

- a change in business strategy;
- internal load changes;
- major new design projects;
- new production equipment investment plans;
- supplier quality and delivery ratings;

* The launch presentation is held in the format of the subsequent review workshops to inform the management team of the detailed project plan

- supplier capacity changes;
- supplier price changes.

There may be others of importance in the particular environment of the business in question, and in setting up a maintenance procedure the strategy formulation process should be critically reviewed for all the relevant factors.

By maintaining a watching brief on these factors, and a record of how the current strategy was devised, the company is in a good position to respond rapidly to changing conditions. The consequences for the business can be assessed, and if necesssary a revision of the strategy carried out.

Application of this maintenance procedure is most effective if linked to the

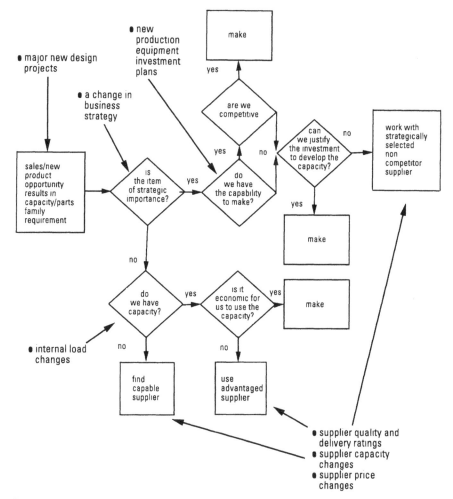

*Fig.5.4 Strategy maintenance check points in the make or buy procedure*

traditional make or buy decision tree which represents the normal outworking of the strategy; such a tree was shown in Fig.2.3. Every time the tree is used, there should be check points which ask whether any of the factors listed above have changed since the formulation of the strategy. This involves keeping a record of these critical formative criteria, and maintaining a live record of their current state.

The point at which these factors would become relevant is indicated on the modified decision tree shown in Fig.5.4. Change in any one of these factors does not necessarily mean a complete review of the strategy. For example, the effect of supplier price changes may be minimal in terms of effect on strategy. On the other hand new design projects or new production equipment investment plans could have major effects. The first step in the maintenance process is to check the original strategy formulation for the significance of the particular factor that has changed. The need for strategy review is then determined by the likely impact of the change in the factor concerned. A key benefit of the systematic approach to make or buy strategy described in this book is the possibility of revisiting the conditions under which it was developed, and the explicit record it provides of the determining factors.

As an alternative to the assessment of these checkpoints everytime the tree is used, it may be advantageous to operate a routine monitoring system. In this case the strategy is kept up to date at all times and the tree ready for use whenever the need arises. In a business where there are very frequent make or buy decisions to be made, it is advisable to operate the maintain routine separately from the use of the decision tree.

## Chapter 6
# Practical experience from past cases

## 6.1 Introduction

The approach to developing a make or buy strategy described in this book has been used in many different businesses. It has been found to work effectively under varying conditions and with varying degrees of expert guidance.

Key aspects from four recent applications are described in this chapter in order to give examples of the range of business issues which can be encountered and to illustrate some of the important consequences for the businesses. The companies concerned all had different reasons for embarking on a make or buy review, but all faced business difficulties of one kind or another and were looking for significant business improvement. Although the identity of the companies is disguised for reasons of confidentiality, the conditions described are all genuine.

Table 6.1 gives the principal dimensions of the companies before the make or buy project. The consequences for each of them are discussed below.

*Table 6.1 Make or buy cases - company characteristics*

|  | Company A | Company B | Company C | Company D |
|---|---|---|---|---|
| Industry sector/ products | large diesel fuel injection equipment | scientific and laboratory instruments | automotive electrical switch controls | special transport containers |
| Turnover | £13m | £100m | £27m | £9m |
| Employees | 337 | 700 | 1200 | 120 |
| Number of manufacturing technologies | 23 | 30 | 15 | 28 |

## 6.2 Four case studies

### *Company A*

This company faced a number of difficult challenges. An ageing population of manufacturing machines required imminent renewal, difficult market conditions had seriously reduced profitability, and the site lease could not be renewed. Closing the business was a serious option.

In this situation the site manager used the make or buy project as a vehicle for business regeneration, and took the courageous decision to commit to not making any of the workforce redundant as a result of the exercise. In the event the apparent risk of this decision was rewarded, and within a year the business had shown significant improvement, with growth in both turnover and profitability, while maintaining employment levels. The entrepreneurial spirit engendered by the project resulted in the identification of hitherto unexploited aftermarket opportunities, and an ongoing programme of supplier selection and development. The business figures showing the effect of these trends are given in Table 6.2 below.

The reduction in shopfloor machines has been achieved by concentrating on the strategically important processes, and eliminating old, noncapable equipment. These were items of plant which were unable to meet current process capability standards without frequent manual intervention.

A number of key aspects stand out from this project.

*1 The value of strong leadership and top management commitment*

The courage to take the plunge and stick to it was a very important factor.

*2 The power of the competitiveness/importance matrix in focusing attention on the few critically important areas*

The matrix for the manufacturing technologies is shown in Fig. 6.1.

*Table 6.2 Make or buy business figures for Company A*

|  | Year 1 | Year 2 | Year 3 |
| --- | --- | --- | --- |
| Sales turnover | £13m | £18m | £21m |
| Headcount | 337 | 344 | 344 |
| Shopfloor machines | 480 | 290 | 180 |
| Floorspace | 200,000 | 145,000 | 110,00 |
| required (sq ft) |  |  | (new factory) |
| Bought out %age |  |  |  |
| of costs | 24% | 43% | 50% |
| Cost savings |  | 15% | 20% |
|  | — |  |  |

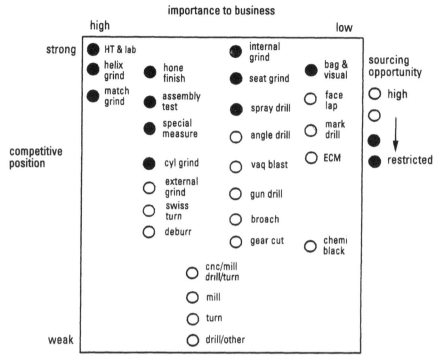

*Fig. 6.1 The competitiveness/importance matrix for Company A*

Fortunately for the business, there were a number of technologies in the top left hand corner, i.e. very important to the business but also very competitive. In addition to the matrix positioning, an assessment of the sourcing opportunity outside the business was made. This is also indicated on the matrix, and provided valuable additional information when prioritising the make-in/buyout options for the manufacturing technologies.

The assessment of the product architecture produced equally significant results. An analysis of the one hundred and thirty components within the product structure revealed that only six of them were of strategic importance. A decision was made to restructure the business around the critical manufacturing technologies and components.

## 3 The benefits of a new approach to sourcing strategy

Initially, as a result of the make or buy review, the business decided to outsource most of the soft stage machining processes, where no significant advantage was available from in house manufacturing. This process of approaching the supply market with a new perspective ultimately generated some new opportunities for business improvement in purchasing practices themselves.

An example of this was in the area of raw material supply. A new supplier of special steels was selected which, although charging a slightly higher price per tonne, was prepared to deliver in small quantities suited to the production schedule and hold stock on behalf of the business. Thus an apparent increase in price was in effect an overall cost reduction. This philosophy of critically considering the total cost of acquisition has been applied throughout the buying activity.

The current situation for Company A is one of transformed business fortunes. It is now located in a new factory, operating a reduced but rejuvenated range of plant and equipment. New attitudes to supplier selection and development have been introduced, and a culture of continuous search for improvement has been instigated. The make or buy review proved to be the carrier for this fundamental business transformation.

## Company B

This scientific instrument maker had an enviable reputation for the technical excellence of its products. It had achieved a position of market leadership, selling at the top end of the price range. Unfortunately, competition from Japan had come into the market selling good quality products at lower prices and there had been a serious decline in profits. Formerly owned and managed by family members whose main interest was the technology of the instrumentation, the business had been sold to a conglomerate. The make or buy review was undertaken as part of a complete reappraisal of the business and the formulation of a recovery plan.

Analysis of product and process technologies revealed a very wide range of technology within the business. This was necessary in the case of product technology, as leading edge expertise in this area gave the products their world beating functionality. The analytical capability of the instrumentation was dependent on this exceptional, and in some cases proprietary, technology. However, many of the process technologies exhibited no such importance to the business, and the company had considerable investment in manufacturing capacity which gave it no competitive advantage.

The process of analysis which positioned the manufacturing technologies on the matrix was unusually detailed in this case. This arose out of two distinct factors. First, people in the company found it difficult to come to terms with the fact that much of the manufacturing capacity could be outsourced with advantage. It required several passes at the analysis to convince them that this was the case. Secondly, the managers in the company were in general analytical scientists, and were conditioned to expect detail and rigour in any process of calculation. This proved useful in terms of developing this aspect of the make or buy methodology, and it was interesting to note that in general successive passes at the analysis aimed at being more precise did not alter the relative position of the technologies on the matrix to any great extent. This is an illustration of the Pareto principle described in Chapter 3, i.e. that there is a law of diminishing returns in terms of the additional information provided by increased data and analysis.

Realisation of the need to concentrate effort on critical processes and activities was a major shift for a company which had hitherto been very highly vertically integrated. As a result of the make or buy review, which lasted about twelve weeks, further programmes of strategic sourcing, or supplier selection, were initiated. These continued intensively for some two years after the initial reorganisation, and thereafter as an ongoing process at a maintenance level.

Definition of the technologies in use by a business is usually a creative process in itself. At first count in this business, some thirty one manufacturing technologies were identified. For the purposes of initial analysis and comparison with the outside world, this was simplified by combination to twenty one technologies. The Pareto principle was then applied and only ten technologies were selected for detailed competitiveness/importance analysis.

The project had significant bottom line impact, resulting from a 15% reduction in manufacturing cost and a 20% reduction in the workforce. The break even point of the business was reduced from 78% of nominal capacity to 48%, with the associated increased resilience to economic up and down swings.

## Company C

Key competitive criteria for this automotive component supplier were cost and the ability to introduce new products quickly. The factory was heavily dependent on the manual skills of a large workforce, but this had the advantage of being relatively low cost. The make or buy project was part of an effort to identify the manufacturing technologies required for the future, while aiming for continuous cost reduction. The team based approach employed by the methodology successfully integrated the factors affecting new product development into the formulation of a new make or buy strategy.

The background of the business was one of being highly vertically integrated, with nearly all possible processes being carried out in house. The management believed that there was a business case to be made for concentrating resources on product design, thermo plastic moulding and assembly, particularly given the potential to reduce the number of sites on which the business operated.

The review justified this expectation, with a number of commodity processes and components emerging from the analysis as good candidates for outsourcing. These included cable forming, presswork and plating. Follow on sourcing projects were started as a result of the make or buy analysis, and the full outworking of the strategy took a further two years before location savings could be fully realised. This illustrates an aspect that we have encountered in many such projects; i.e. that management expectations of the timescale over which such a significant shift in strategy can be achieved are usually overoptimistic. The process of moving manufacturing technologies in and out of the business takes months if not years, given the degree to which they are integrated into the human and infrastructural systems. Further, even when the move has been made, and new suppliers have been identified, there is a serious time commitment to building up the new supplier relationships. The impact on the role of purchasing and the calibre of purchasing personnel is also often overlooked.

## *Company D*

This is an entrepreneurial family owned business with a small share of the specialised vehicle container market. Having successfully introduced MRP into the business and gained ISO9000 approval, the company was looking for a successive stage of business challenge and improvement. The ideas of World Class Manufacturing appeared to offer a way forward, and the company joined a local TEC (Training and Enterprise Council) programme which majored on this theme. A key part of the programme was a make or buy strategy review, using the methodology described in this book.

The unusual aspect of this project was that there was minimal outside involvement. An external consultant familiar with the methodology provided some initial training and facilitated a number of key sessions, but for the most part the company carried out the analysis and implementation for itself. The results have been significant in terms of taking costs of some £400 000 out of the business which has a turnover of about £10m, and enabling it to weather a period of difficult trading. The market is now picking up, and the company expects to return to secure profitability. This will be on the basis of the restructuring of the manufacturing processes and the supply base that has followed the make or buy review.

A number of key learning points, resulting in business improvement, emerged from this project for the company. They arise from the fact that the derivation of the competitiveness/importance matrix is a powerful and evolving process.

*1 Apparently core manufacturing processes may be outsourced with advantage*

The first pass at the matrix was drawn up during a one day workshop facilitated by the external consultant, and is shown in Fig.6.2. Interest focused on several of the key manufacturing processes, for example the manufacture of the dished ends to the containers which were critical in terms of product quality. An important external input here was to challenge the internal perception that the business was fairly competitive in this process. This led to a review of what customers really valued, which in turn led to some significant changes of view when actual current customer opinion was tested.

In particular, the company had been proud of the polished internal finish that it was achieving in the containers. This was necessary partly due to limitations of the in house processes for forming the dished ends, and also partly because the company believed this was what customers wanted. In fact, on further discussion with customers, it emerged that they preferred the surface of the plate to be in the untreated form as delivered from the steel mill. In this context, any improvement in build accuracy and reduction of welding, would be valued by the customer.

Revision of the matrix in the light of better information resulted in a

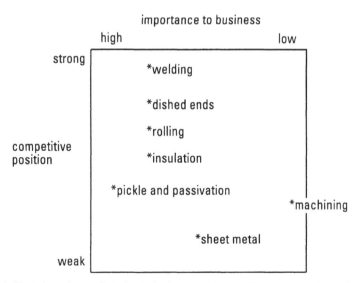

Fig. 6.2 First view of manufacturing technology competiveness/importance matrix after a one day workshop

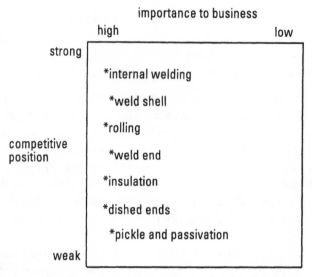

*Fig. 6.3 Revised manufacturing technology/importance matrix-selected techniques*

subdivision of the welding processes which were of varying performance. There was a realisation that some of the manufacturing technologies were more important than first thought, and that in general the company's competitive postion with these was weaker than the first assessment suggested. The revised matrix is shown in Fig.6.3.

This process of establishing and validating the matrix positioning led ultimately to the external manufacture of dished ends, with resulting fixed and variable cost savings and improvements in build quality.

## 2 Matrix analysis drives improvement to internal manufacturing processes

The increased importance of weld quality, given the customers' desire not to have the internal finish polished, resulted in the development of new welding processes with improved control and accuracy. These are innovative processes in which the company is integrating technologies new to the factory to create a novel solution tailored to its own needs.

## 3 Buying procedures are scrutinised

A complete review of steel buying, stocking, handling and sizing resulted from the critical analysis generated by the project. It was found that steel accounted for 25% of the bought out costs, and that it was purchased in 400 varieties due to specification of nonstandard sizes. Analysis of the manufacturing processes simultaneously with buying procedures enabled changes in both to work together for mutual advantage. Improvements to manufacturing accuracy meant that a standardised steel buying policy could be adopted, which eliminated most of the variety. As a result stock value has been reduced from £450 000 to zero, and significant annual savings have been generated.

Naturally such changes are not effected without some internal resistance, typically from people involved in production and quality assurance. However, the benefits of the changes have been seen in both the short term (contribution of £400 000 to the cash flow) and the long term, since the company is now in a good position to take advantage of a market upturn.

## Chapter 7

# Useful tools and techniques for the project team

## 7.1 Introduction

There are many aspects of the work of the project team devising the new make or buy strategy that will require considerable thought and analysis. It has been found helpful for the team to have a range of analytical tools and techniques at its disposal. These can assist in sharpening the issues for discussion, present data and information in a way which facilitates decision making or simply act as vehicles whereby the team can work together and share conclusions. A range of these techniques will be described, although this list is by no means exclusive and there may be many more which are equally appropriate and useful.

## 7.2 Nine techniques

### 1 Business scope (users/needs matrix)

In addition to the general questions given in step 1 of Chapter 4, the matrix of users/needs can give powerful insights into the way in which our products satisfy customers. It also indicates areas where we may have overlooked a market opportunity. An example is shown in Fig.7.1.

The initial step is to make two lists: *whose needs do we satisfy?* and *what needs do we satisfy?*

These two dimensions of users and needs are then assessed against each other by means of the matrix shown in Fig. 7.1. The intersection points of the matrix can be marked simply by a cross, or by a figure showing the size of that particular market. Where intersections are not marked, the question is whether that point is a missed opportunity. Fig.7.2 gives an example completed users/needs matrix for a scientific instrument company.

### 2 Key success factors

The KSFs for a manufacturing business will be variations on the theme of

| needs / users | satisfied | nonsatisfied |
|---|---|---|
| | needs we satisfy at present | needs we may (want to) satisfy in the future |
| **actual** — people or organisations whose needs we satisfy at present | | |
| **potential** — people or organisations whose needs we may (want to) satisfy in the future | | |

*Fig. 7.1 The users/needs matrix*

| | | satisfied | | | | nonsatisfied | |
|---|---|---|---|---|---|---|---|
| **needs / users** | | measuring | | analysis | | measuring | |
| | | sound | vibration | gas | pollution | gas flow | pressure |
| **actual** | manufacturing industry | 5 | 15 | 8 | 5 | ● | ● |
| | consulting engineers | 2 | 4 | | | ● | |
| | government/ regulators | 1 | | 2 | 6 | | ● |
| **potential** | utilities | | | | | ● | ● |
| | construction industry | | | | | | ● |

*Fig. 7.2 Users/needs marix for a scientific instrument company*

cost, quality, delivery and flexibility. It is important to realise that it may not be possible to define these at a business level, and that to arrive at something which we can operationalise we have to work at product/market group level.

The conclusion may be that the KSFs are largely similar within the business, but that the weighting between them varies with product/market groups. A multidisciplinary team made up of senior representatives from manufacturing, design, marketing and sales should agree what the KSFs are and decide their weighting for different product/market groups. An essential ingredient to this process is the customers' viewpoint, since they ultimately define the KSFs by their purchase decisions.

The conclusions of the team deliberations can be recorded on a table such as the one shown in Table 7.1. Note that the sum of each product/market group weighting should be unity.

| KSF | Product/market 1 | Product/market 2 | Product/market 3 |
|---|---|---|---|
| Quality | 0.15 | 0.1 | 0.2 |
| Features and performance | 0.25 | 0.4 | 0.25 |
| Price | 0.4 | 0.2 | 0.25 |
| PIP lead time | 0.05 | 0.1 | 0.2 |
| Delivery performance | 0.1 | 0.05 | 0.05 |
| Responsiveness | 0.05 | 0.15 | 0.05 |

*Table 7.1: KSF weighting by product/market group*
PIP = product introduction process

## 3 Business strategic issues

The business strategic issues are the decision areas that are going to require significant top management attention in terms of the future direction of the business. The heart of the process to define these is a SWOT analysis (strengths, weaknesses, opportunities, threats) which is carried out by means of an executive workshop. The steps in the process are as follows:

### (i) Brainstorm the SWOTs

The idea is to compile as many of these as possible as short words or phrases. A clear definition of strengths and weaknesses as internal characteristics of our organisation, and opportunities and threats as external trends or conditions which all competitors face alike, is essential.

## (ii) Simplify the SWOTs

Having gathered all relevant ideas for each category in turn, the next step is to group the ideas into natural clusters around a theme. For example, it is normal to find that some twenty five individual ideas of strengths in our organisation will summarise down to about five significant clusters of ideas, each of which can be expressed as a short sentence.

## (iii) Confrontation of the SWOTs

The impact of the strengths and weaknesses on the opportunities and threats facing the company is evaluated by means of a confrontation matrix, Fig. 7.3.

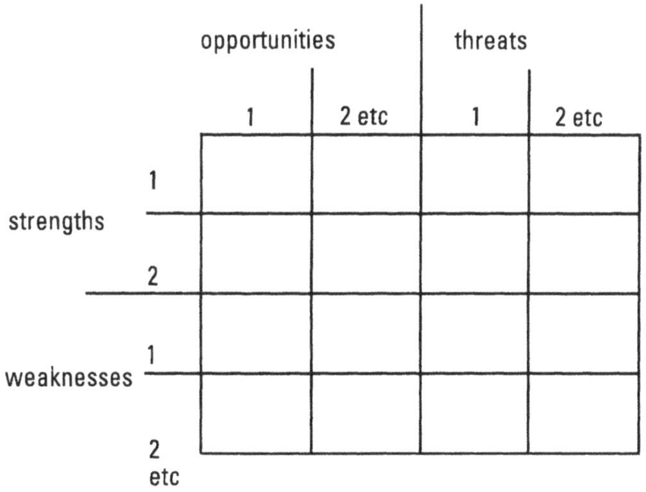

*Fig. 7.3 SWOT confrontation matrix*

Each confrontation can be scored against the others, with regard to the guidelines given in Fig.7.4. The scores range from 0, no effect at all, to 4, very great effect. The scoring process is carried out by each member of the workshop, and the individual scores are then totalled. This is in effect a voting process which evens out any extreme views. However, any wide diversity of judgement should be revisited to check common understanding. Intersections on the matrix which show a total high score are points of

|  | opportunities | threats |
|---|---|---|
| strengths | how much does this strength help us to exploit this opportunity? | how much does this strength help us to combat this threat? |
| weaknesses | how much does this weakness hinder us in exploiting this opportunity? | how much does this weakness hinder us in combating this threat? |

*Fig. 7.4 SWOT scoring guidelines*

significant interaction between strengths, weaknesses, opportunities and threats. These points are highlighted on a matrix showing the total scores.

## (iv) Deriving the strategic issues

The total matrix is examined for patterns of significant interaction. It is these interactions which can be expressed as the strategic issues facing the business. It may be that while we have strengths which would help us to exploit a particular opportunity, a major weakness is holding us back. The significant interactions are extracted from the matrix in this way and verbalised as the strategic issues. An example is shown in Fig. 7.5.

| | | opportunities | | | threats | |
|---|---|---|---|---|---|---|
| | | op1 | op 2 | op 3 | thr 1 | thr 2 |
| strengths | str 1 | | + | + | | |
| | str 2 | + | | | | + |
| weaknesses | wks 1 | | | | | - |
| | wks 2 | | | - | - | |
| | wks 3 | - | - | - | | |

*Fig. 7.5 Identifying patterns of significant interaction*

The careful formulation of these strategic issues is important because later in the process the impact of various technology choices on these issues will play a large part in choosing the optimal strategy. Ideally no more than four or five significant strategic issues should be extracted from this matrix, and they should be described in a manner which is readily understood by everyone in the business. In this way they become an important part of the communication process linked to the make or buy review, and provide a focus for driving change.

The plus signs indicate the positive effect of strengths and the minus signs indicate the negative effect of weaknesses. The strategic issues are indicated by clusters of significant interaction drawn either vertically or horizontally on the matrix.

## *4 Business activity assessment*

The central analysis of this make or buy methodology is of the manufacturing technologies and product architecture in terms of importance and competitiveness. However, a necessary precursor to deciding where to invest in technologies is an appraisal of the manufacturing function as compared with the entire spectrum of business activities. It is particularly important to carry out this step if it is not clear that manufacturing is the focus of the problem facing the business; it could be that one of the other activities would give a better return for the project effort about to be expended. In order to make this assessment six key activities can be considered:

marketing
design and development
acquisition and sourcing
manufacturing
distribution
after sales support.

The technique of determining importance and competitiveness via links to the KSFs, described in step 8, is used. Similarly, there will be a step in the analysis to identify the relevant output measures from each business activity which relate to the various product/market KSFs in technique 2.

The output from this exercise should be a matrix with all six activities plotted, see Fig. 7.6. From this matrix the standing of manufacturing with regard to the other activities can be assessed. It is at this point that the decision is made whether or not the make or buy project is worthwhile, given the importance of manufacturing to the business as a whole. The investment of approximately three weeks, effort in answering this question can save many weeks of subsequent work of limited value, if manufacturing issues are not central to the company's business problems.

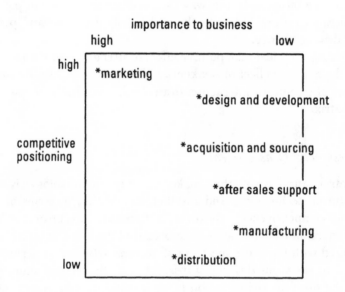

*Fig.7.6 The business activities matrix*

## 5 Process quality management (PQM)

This is essentially an intensive team building activity, which is primarily designed for the top management team of the company. However, there are some specific outputs of the process which could be useful to the make or buy project. The key requirements are full commitment from the management team leader, involvement of all the management team and time away from day to day pressures to complete the process. Outputs include agreement on the company mission, determination of the key success factors and a description of the main business processes which affect the performance as measured by the key success factors. A full account of the process is given in the *Harvard Business Review* of November-December 1987, under the title 'Getting things done: how to make a team work'[17].

PQM goes further into some aspects of company performance than we require for make or buy strategy formulation. However, in a situation where there is little management team concensus and manufacturing is not the main focus of the business problems, it can be a very useful tool.

## 6 DTI competitive manufacturing workbook[18]

Providing a means of auditing the current manufacturing strategy of a business, this workbook is a well tested front end to the manufacturing strategy formulation process, and as such has a wider scope than the make or buy strategy. It makes an assessment of how well the existing manufacturing strategy matches the business objectives of the company. The methodology is built around seven worksheets and is normally carried out by a group of people in the business. It could be particularly useful in a company where manufacturing is the focus of the business issues, but some diagnostic work is necessary to determine the scope of the required improvement project. It could be used in a short form in order to validate the view that make or buy aspects are the likely focus of further strategy development. For further details see the workbook itself.

## 7 Value chain analysis

Mapping of the business processes of product delivery, from obtaining raw material to delivery to the end customer, is provided by value chain analysis. It gives a picture of how much of the total process is within our control. It can show up opportunities to apply capabilities to other parts of the business process (value chain), and by comparison with the competition show where we may be vulnerable. It is possible to carry out this type of analysis at industry level to get a feel for the big picture, or at a more detailed level within our own company to get an understanding of where value is added in the manufacturing process, as already shown in Fig. 2.6 Both levels of analysis can be useful in formulating make or buy strategy, since they help to identify opportunities to move activities in or out of the firm.

## 8 Brainstorming

Brainstorming is a group process for problem solving which works by encouraging creativity and the flow of ideas. The problem to be addressed is first presented and generally discussed to make sure that all participants have a clear idea of what is required. The process then moves into a phase where all possible solutions are suggested by all members of the group and recorded in a totally noncritical atmosphere. Any idea, no matter how unlikely, is welcomed and recorded. In this way unusual and original ideas often emerge, and creative solutions can be triggered by the association of uncoventional ideas.

It is helpful to get the group into a relaxed frame of mind before the brainstorming session, by means of a warm-up exercise or even an enjoyable social event such as going for a drink or a meal. While it is often an advantage to have a mix of people from different backgounds or functions in the business in order to access a wide range of ideas, care must be taken to avoid the inhibitions which may follow from having the boss present! This does not have to be a problem as long as everyone realises that any idea is as good as another, and everyone is entitled to have their say. The session itself should take no more than an hour, and if it is felt that there is scope to do more or tackle a related problem, then it is better to reconvene at another time when people have mentally recharged.

## 9 Quality function deployment (QFD)

Quality function deployment is a structured approach to product and process design that is driven by customer requirements. In this respect it has similarities with the ideas behind the make or buy review, and it may be useful to apply QFD in the latter parts of the strategy development process, when the detailed sourcing options for manufacturing technologies and parts families are under review.

QFD is applied in four stages:

stage 1 — translates customer requirements into design requirements
stage 2 — translates design requirements into critical part characteristics
stage 3 — translates critical part characteristics into critical process parameters
stage 4 — translates critical process parameters into production requirements.

The schematic of the QFD process is shown in Fig. 7.7.

Application of QFD usually requires some experienced facilitation for first time users of the technique. Its use may be regarded as a learning process in itself, and the methodology can be explored without the help of an expert, as long as there is no short term requirement for an output.

The value of the QFD methodology lies in its rigorous and systematic approach, which can be used as an independent validation of the manufacturing technologies and product architecture components which emerge from the make or buy analysis as important.

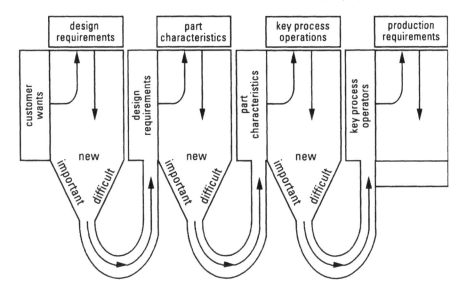

*Fig. 7.7 Schematic overview of the information flow through the four phases of QFD*

## Chapter 8

# Where to go from here

## 8.1 Do we really need to make any change?

It is an unfortunate fact of life that most companies seem only to address fundamental issues such as make or buy strategy when facing an imminent crisis. It appears that many management teams make the assumption that their business will survive and possibly even thrive without any major interventions. The difficulty with this position is that deterioration in the business situation is often imperceptible, and the loss of market share, slippage compared with competitors or detachment from actual customer requirements may only be realised too late. The analogy of the frog sitting happily in a pan of water, being slowly heated up until it is boiled alive, springs to mind. The grin on its face seems strangely at odds with the circumstances to the outside observer!

How much better to anticipate changing circumstances, by maintaining an ongoing assessment of a company's match to its environment. The make or buy strategy formulation, together with the maintenance of the strategy is one way to do this. You can start by asking a few simple questions of your current practice.

First ask yourself:

- is make or buy an explicit part of our manufacturing strategy?

and if *yes*:

- do we have make or buy decision making routines in place that are driven by the strategy?

If the answer is *no* to either, then you should think seriously about carrying out a make or buy review. Even if you have answered *yes*, you may not be on safe ground unless you are able to answer the following:

- how much of your sales value do you buy from suppliers?
- how does this compare with your key competitors?
- is profit sufficient to enable you to invest for growth?
- are you in a cyclical market place?

- is your level of fixed cost a problem?
- how does your manufacturing capability compare with competitors?

  - cost, quality, delivery
  - age and performance of plant

- do you know which manufacturing processes and parts of the product architecture have a critical effect on satisfying the customer?

If you can answer all these questions and are comfortable with the answers, then in all probability the make or buy strategy of your business will be on a fairly sound footing. However, if you cannot answer with confidence then it could be advantageous to carry out some initial diagnostic checks before considering a full make or buy review.

## 8.2 How to get started

You can start to develop a new make or buy strategy right away by critically examining the role of manufacturing in the context of your whole business. There are many techniques available to do this, such as the business activity assessment in Chapter 7, or by means of the methodology in the DTI publication *Competitive manufacturing*.

This will make clear the relationship between customer requirements and the necessary structure and performance of your manufacturing operations. The next step is to review your manufacturing processes and the structure of your main products in terms of the contribution these make to the satisfying customers. The important areas that are identified by this review are the areas on which to concentrate your product and process capability. These diagnostic steps will confirm whether make or buy is the area for further work, or whether some other aspect of the manufacturing strategy is the priority.

It may be fairly clear from trying to answer the initial questions above, that make or buy issues are the central concern for the company, in which case the methodology outlined in this book can be started right away. The methodology is complete in itself and does not require any prework, given that commitment has been made to make or buy as the focus of interest.

## 8.3 Top management commitment

A critical factor to get right from the start of the project is senior management support. This is needed to endorse the objectives of the make or buy review, and also to ensure that the necessary resources to carry it out will be available. This is not usually a problem if the managing director has initiated the work, but if a more junior member of the organisation is championing the need for a review then securing this top team support is an early requirement. It may be advisable to carry out the first diagnostic stages in order to have convincing evidence, but top level support will certainly need to be in place before detailed work starts.

## 8.4 Other sources of practical help

There are a number of sources of further help with the practical application of the ideas presented in this book. They include other work on related topics, a short course based on the approach and facilitation and consulting around the methodology by people experienced in its use. Here are the most relevant:

A short booklet [1] to raise management awareness of make or buy issues has been published by the DTI as part of the Managing in the 90s programme. Entitled *Make or buy: your route to improved manufacturing performance?*, it provides good preparatory material for a management team (and others in the business) thinking of embarking on a make or buy review. Copies of the booklet can be obtained from DTI Publications: Tel: 0171 510 0144, Fax: 0171 510 0197

The Cambridge Programme for Industry runs a one-day course on make or buy issues based on the content of this book. Entitled 'Make or buy ? Your route to better business and manufacturing performance', details are available from:

University of Cambridge Programme for Industry
1 Trumpington Street
Cambridge CB2 1QA
Tel: 01223 332722, Fax: 01223 301122

Consultancy support for the application of the make or buy approach in this book is available from the Manufacturing Business Unit of CSC Computer Sciences Ltd. Their staff have been involved with many different business applications and can be contacted at:

CSC Computer Sciences Ltd
PO Box 52

Shirley, Solihull
West Midlands B90 4JJ
Tel: 0121 627 4040, Fax: 0121 627 3704

Many practical tools and techniques to support business improvement project teams (some of which are used in this make or buy approach) are described in the *CSC manufacturing industry handbook*, also known as the *Mini Guide*. This is available from the CSC address above.

The assessment of your current manufacturing strategy is the subject of the book *Competitive manufacturing*. This can help to identify make or buy as the central issue of concern for a manufactuirng business. It is available from:

IFS Publications,
35 - 39 High Street,
Kempston,
Bedford MK42 7BT

Extracts from the report by the Engineering and Manufacturing Committee of the Society of British Aerospace Companies on *Make or buy policies within the aerospace sector* are available free of charge from Michelle Stafford at the Society:

Society of British Aerospace Companies
60 Petty France
London SW1H 9EU
Tel: 0171 227 1024, Fax: 0171 227 1025

The issues of customer/supplier relationships have been the subject of research by the University of Glasgow. The results of this work are available in a self help form from:

Supply Chain Management Group
59 Southpark Avenue
Glasgow G12 8LF
Tel: 0141 330 5696, Fax: 0141 330 5698

Facilitation with the application of the approach described in this book is available from people involved in its development. Contact:

Practicam
20 Trumpington Street
Cambridge CB2 1QA
Tel: 01223 338187, Fax: 01223 332797

# References

1. PROBERT, D.R. (1995): 'Make or buy: your route to improved manufacturing performance?' Department of Trade and Industry, London
2. CULLITON, J.W. (1942): 'Make or buy'. Business Research Study 27, Harvard University Graduate Business School
3. GAMBINO, A.J. (1980): 'The make or buy decision' (National Association of Accountants, New York)
4. SBAC (1995): 'Make or buy: The decision' Department of Trade and Industry, London
5. SKINNER, W. (1974): 'Manufacturing: missing link in corporate strategy,' *Harvard Business Review*, May-June
6. HAYES, R.H., WHEELWRIGHT, S.C. and CLARK, K. (1988): 'Dynamic manufacturing: creating the learning organisation' (Free Press)
7. MACBETH, D.K., BAXTER, L.F., FERGUSON, N. and NEIL, G.C. (1990): 'Customer supplier relationship audit' (IFS)
8. SAKO, M. (1992): 'Prices, quality and trust: interfirm relations in Britain and Japan' (Cambridge University Press)
9. WILLIAMSON, O.E. (1975): 'Markets and hierarchies: analysis and antitrust implications' (Free Press)
10. MONTEVERDE, K. and TEECE, D.J. (1992): 'Supplier switching costs and vertical integration in the automotive industry', *Bell Journal of Economics*, 13
11. BUZZELL, R.D. and GALE, B.T. (1987): 'The PIMS principles: linking strategy to performance' (Free Press)
12. PORTER, M.E. (1980): 'Competitive strategy: techniques for analysing industries and competitors' (Free Press)
13. HARRIGAN, K.R. (1983): 'Strategies for vertical integration' (Lexington Books)
14. PRAHALAD, C.K. and HAMEL, G. (1990): 'The core competence of the corporation', *Harvard Business Review*, May-June
15. ABETTI, P.A. (1989): 'Linking technology, business strategy' (The President's Association, American Management Association, New York)

16. HAYES, R.H. and WHEELWRIGHT, S.C. (1984): 'Restoring our competitive edge: competing through manufacturing' (Wiley)

17. HARDAKER, K. and WARD, B.K. (1987): 'Getting things done: how to make a team work,' *Harvard Business Review*, November-December

18. PLATTS, K.W. and GREGORY, M.J. (1989): 'Competitive manufacturing: a practical approach to developing a manufacturing strategy' (IFS)

# Index

# A Comprehensive Approach to Neutron Diffraction